Building for the Atomic Age

Building for the Atomic Age

An industrial archaeology of the United Kingdom's nuclear industry

Wayne D. Cocroft

Historic England

Published by Liverpool University Press on behalf of Historic England, The Engine House,
Fire Fly Avenue, Swindon SN2 2EH
www.HistoricEngland.org.uk

Historic England is a Government service championing England's heritage and giving expert, constructive advice.

First published 2026

ISBN: 978-1-83624-577-3

British Library Cataloguing in Publication data
A CIP catalogue record for this book is available from the British Library.

Typeset in Georgia Pro Light 9.5/11.75

Page layout by Carnegie Book Production

Printed in the Czech Republic via Akcent Media Limited

The manufacturer's authorised representative in the EU for product safety is:
Easy Access System Europe, Mustamäe tee 50, 10621 Tallinn, Estonia
https://easproject.com (gpsr.requests@easproject.com)

Front cover: Sizewell, Suffolk. [© Historic England Archive, 26725_019]
Back cover: UKAEA Culham, Oxfordshire, JET Assembly Hall, full size training model of the tokamak. [© Historic England Archive DP438105]
Frontispiece: Sellafield, Cumbria, is one of largest and most complex nuclear industrial facilities in the world.
Photograph taken in 1995. [© Historic England Archive (Aerofilms Collection), EAW650529]

Contents

In the early 1950s, in his vicarage study to the north of Oxford in the quintessentially English village of Steeple Barton, the landscape historian William Hoskins worked on his pioneering book *The Making of the English Landscape*. His verdict on the 20th century's impact on the English landscape was harsh, uglifying and destroying its meaning. His annoyance was perhaps reinforced by the noisy intrusion of American atomic bombers on their approach to RAF Upper Heyford 5km to the east. He lamented the loss of country houses and their surrounding parkland, but noted how some, perhaps thinking of the Harwell research centre to the south, had been taken over 'by what some villagers describe detachedly as 'the atom men', something remote from the rest of us'.[1] While this view of otherness may have characterised the beginnings of the atomic industry, as it matured, the aspiration of its senior managers was to integrate the workforce into local communities. As this study will also show, contrary to Hoskins' view, a reoccurring and characteristic theme of the British nuclear industry was its concern to ensure its facilities made a positive contribution to their local landscape.

W. G. Hoskins 1955 *The Making of the English Landscape* (Harmondsworth: Penguin) 1978 rpt

1 Hoskins 1955, 298.

Preface

This book is concerned with the material culture of the United Kingdom's nuclear industry and the interconnectivity of its technology, architecture of its structures and impacts on wider landscapes. It begins with a brief history of early atomic research, often conducted in pre-existing university buildings. Experimentation quickly outgrew the research bench, and increasingly larger purpose-built laboratories were required, with complex and potentially hazardous experimental machinery supplied by the heavy electrical engineering industry. During the late 1930s, British scientists, and many fleeing persecution on the continent, were amongst the first to recognise the potential of atomic energy both to create fearsome weapons and as a potential source of power. At the beginning of the Second World War, anxieties about German activities in this field led the government to fund university academics and experts from private industry to begin tentative steps to building an atomic bomb.

Scarcity of resources and vulnerability from aerial attack moved the focus of research to the United States and, to a lesser extent, Canada, with many British and recently naturalised scientists assigned to the American Manhattan atomic bomb project. Immediately after the war, Britain embarked on a programme of atomic research, with government-backed research establishments given an initial goal to build an atomic bomb. This work also relied heavily on the support of the existing chemical and engineering firms, but new industries with a wide geographic distribution also grew to supply fissile materials and other specialist plant and instruments.

In the early 1950s, attention was turned to the possibilities of power from nuclear energy. From a potentially bewildering choice of reactor configurations in 1955, the United Kingdom announced the world's first civil nuclear power programme. Over the next decade, the first series of uranium-fuelled, graphite-moderated and gas-cooled Magnox reactors came into service, followed by a more powerful fleet of advanced gas-cooled stations. Research also continued into other reactor types; these came to naught, and in the early 1980s, an American-inspired pressurised water reactor design was selected for the last nuclear power station completed in this country.

Seventy years since the construction of Calder Hall power station – the first in the world to be connected to a national grid – and with a new generation of nuclear power stations on the horizon, now is an appropriate moment to explore this industry's past. This work draws its inspiration from field investigations, site visits, archive research and conversations conducted over more than 20 years. It is profusely illustrated by photographs held by the United Kingdom's national heritage agencies and the nuclear industry, including historic images alongside many specially commissioned for this book.

Acknowledgements

This book has had a long gestation, and this is reflected in the debt of thanks owed to many individuals, organisations and institutions. It has its origins in a then English Heritage project, *Strategy for the Historic Industrial Environment Report: England's Atomic Age* (Cocroft 2006). It may also be seen to complement *Cold War: Building for Nuclear Confrontation 1946–89* (Cocroft and Thomas 2003), which contained some discussion of nuclear weapons research sites and provided a discussion of nuclear weapons deployment sites in England.

Over subsequent years the author has had privileged access to many nuclear licensed sites in the United Kingdom and abroad and is grateful to their personnel for facilitating access and explaining their operation. These include staff at Sellafield and at United Kingdom Atomic Energy Authority (UKAEA) Culham, where Nicholas Balshaw provided expert guidance around the Joint European Torus. The work has also greatly benefitted from conversations with members of the Nuclear Decommissioning Authority's (NDA) Heritage Panel, including Norman Bird, Philip Greatorex, James Gunn, Elizabeth Norton and Bob Radford. My long-standing collaborations with Professor John Schofield, University of York, have greatly influenced my approaches to the archaeology of recently past technological worlds. Collaborative fieldwork at the Nevada Test Site with him and colleagues at the Desert Research Institute, Las Vegas, revealed the archaeological traces of this contested place, including a vast desert peace camp and, within the fence, material legacies of the United Kingdom's overseas testing programmes. Adrian Armishaw, Sywell Aviation Museum, kindly provided access for photography of objects in its care.

This book has also been made possible by the support of many colleagues at Historic England. These include the heads of industrial heritage Keith Falconer and Shane Gould. Magnus Alexander, Sarah Newsome, Elaine Jamieson and Jonathan Millward contributed to fieldwork at Orford Ness. Sarah also participated in work at Foulness, where Stephen Miles also assisted, in addition to helpfully providing copies of many obscure technical journals. Nicola Cryer, Historic England librarian, has greatly assisted with the speedy delivery of journal articles. Many other members of staff in the Historic England Archive have efficiently retrieved archival material and supplied images used in the book. Fieldwork on nuclear weapons deployment sites has been made more informative and enjoyable through the input of Roger J. C. Thomas. Damian Grady and David MacLeod have taken aerial photographs of nuclear facilities in their wider, often designed landscapes. Historic England professional photographers Alun Bull, Steve Cole, James O. Davies, Derek Kendall, Sarah Lever and Christopher Redgrave have brought their skills to document these sometimes utilitarian structures. I am also grateful to previous generations of anonymous professional civil service and commercial photographers whose work now provides an invaluable historic record. John Vallender has significantly enhanced the appearance of the book through his artwork. David Went offered insightful comments on an early draft of the text. North of the border, Miriam McDonald and Miles Oglethorpe, Historic Environment Scotland, have worked to secure the historical documentation of the nuclear industry in Scotland.

The book has also been greatly enhanced through conversations and friendships made through the Charterhouse Nuclear History Group and its predecessors, now convened by Dr Richard Moore. For over two decades this group has brought together scientists, engineers, managers and policy makers from across the nuclear sector, together with historians exploring this topic. Discussions with the late Dr Kate Pyne, formerly historian at the Atomic Weapons Establishment, and Drs Jonathan Aylen and William Nuttall have greatly assisted in piecing together the story and locations involved in the development of Britain's first atomic weapon. Jonathan also kindly acted as an external reviewer of the manuscript. I am also glad for being invited to participate in various Nuclear Cultural Heritage events coordinated by Dr Egle Rindzevičiūtė, which have introduced me to a new generation and international group of researchers working in this field.

I am obliged to the Royal Commission on the Ancient and Historical Monuments of Wales, through their secretary Christopher Catling, for permitting the use of some of their images. I am also grateful to Paul Hetherington, Elizabeth Norton and Simon Tucker for facilitating the use of NDA material and to Malcolm Crook (Magnox), Martin Denvir (Sellafield Ltd) and James Gunn (Nuclear Restoration Services) for the provision of historic photography. The book was brought to production by the copyediting and proofreading skills of Paul Stamper. I also thank Alison Welsby and Sarah Warren at Liverpool University Press and Lola Moreno LaVey at Carnegie Book Production for their support in bringing the book to press.

1

Introduction

This book sets out to explore the industrial archaeology of Britain's nuclear industry. Its primary concern is with the physical manifestations of the industry, its buildings, landscapes and technology. These are set within a historical and social context; parts of the story will be familiar, but many of the locations where these activities took place have only been known to relatively few people. Traditionally, the raw material of industrial archaeology has been the decaying remains of former industries, whether on neglected urban fringes or remote rural areas, where today their associated settlements may appear oddly isolated. Higher environmental expectations and development pressures on brownfield sites now ensure that former industrial land is quickly remediated. For the industrial archaeologists of the future there will be few tangible traces left of modern industrial activity, especially complex technological industries such as the exploitation of nuclear energy.

The practice of archaeology, including building investigation, usually involves close physical inspection, note taking, photography, plan making and analysis. For many practical reasons, particularly health and safety concerns, and, in the first decades of the 21st century, security issues, these conventional methods of investigation have rarely been possible in the exploration of the country's nuclear past. It might be questioned whether the discipline of industrial archaeology has a place in the investigation of late 20th- and early 21st-century industry, given an abundance of written, drawn, photographic and other sources. What defines this study as archaeological is its primary concern with the built heritage and material culture of the industry. However, to provide a broad overview of the industry, it is essential to explore its technical history; for example, the evolution of reactor technology is a crucial factor in determining the outward architectural form of a power station. Industrial archaeology is also concerned with process, production and industrial operations, as they are revealed by machines and building layouts, and the social experience of work and life in associated communities.

The industrial archaeologist of the recent past is fortunate in that there is potentially a huge wealth of illustrative material that may be consulted. As with any historical sources, there are inherent biases within this material, and images of routine day-to-day working life are underrepresented. This is especially true of an industry where private cameras are barred, and the historic record is reliant on technical and posed press images taken by official photographers. Nevertheless, they provide some of the richest sources for illustrating the human experience of working in the industry, the special types of clothing required and the interfaces between people and the machines used to handle some of the deadliest materials known to humanity. At the time of writing, many atomic-related files at The National Archives were retained for review, and records related to the civil industry were in the process of being transferred to the Nucleus archive in Wick (Caithness). This has restricted the range of official documents currently available for study. However, the contemporary trade press, which proudly reported the latest technical developments in the industry, has been an invaluable source.

In other instances, there is a relative paucity of historical material and here traditional archaeological methods have come to the fore. Documentation relating to Britain's early atomic bomb project during the late 1940s is patchy and it is challenging to investigate the networks of researchers and companies who contributed to this enterprise.[1] At the government research establishments at Fort Halstead (Kent) and the Royal Arsenal, Woolwich, through air photographic analysis it has been possible to identify the structures adapted and built to support the work, and physical investigation of the structures was used to explain their functions. At the former weapons testing sites at Foulness (Essex) and Orford Ness (Suffolk), documentation is also partial; here traditional archaeological methods, including topographic survey, air photographic and building analysis, were used to understand the development of these places and something of their past activities.[2] Similar work by Birmingham Archaeology has revealed the history of the pioneering wartime work of the early gaseous diffusion plant at Rhydymwyn (Flintshire).[3] Given the highly secretive nature of these places, photographs taken during these investigations have often provided some of the first publicly accessible images of these sites.

High-tech Britain

Britain's nuclear programme was one of several high-tech industries, including radar, jet engines, electronics and rocketry, which had emerged from wartime military research and where British scientists and engineers were among the leaders in their fields. In the chilling political, economic and military stand-off between East and West that followed the Second World War, their development was vital to national security. It was argued that the deterrent effect of nuclear weapons would spare the population from the horrors of another world war, but as the country recovered from wartime deprivations, these technologies also offered a vision of an industrial renaissance leading to new wealth that might be used to create a better society, adequately housed, freed from the threat of mass unemployment, and cared for by a national health service. A common feature of all these projects, and a legacy of wartime industrial organisation, was a strong link between the state, reflected by government research establishments and their close collaboration with private industry. Typically, after the basic science and technology was developed within the government research establishments, commercial companies were invited to tender for the manufacturing and construction contracts, although increasingly supported by their own private research facilities. A contemporary commentator speaking at the 1958 Public Inquiry into the location of the Trawsfynydd (Gwynedd) nuclear power station commented of the industry's workforce, 'They felt they were in an industry which was one of the coming things – one of the industries which is in the forefront of present development. They felt themselves to be taking part in something really essential and worthwhile not only for their own benefit but the national benefit.'[4]

During the 1950s, 'atomic' became synonymous with everything futuristic and modern (Fig. 1.1). To convey a sense of modernity, the prefix 'atomic' was given to everything from draught excluders to expresso coffee machines. Atomic imagery was taken further: the 1958 Brussels World Trade Fair was dominated by the Atomium, a 102m structure representing an iron crystal, comprising nine 18m spheres joined by staircases and escalators. On a smaller scale, umbrella stands, coat hooks, magazine racks and a host of other household items were fashioned from wire capped with coloured balls that mimicked scientific atomic models. Mastery of the atom was a matter of national pride, and internationally the message was widely conveyed on contemporary postage stamps. Beyond these cultural markers, through the global spread of radioactive traces, the atomic age may also provide one of the 'golden spikes' defining the Anthropocene, a geological era induced by human industrial activity.[5]

Across the United Kingdom the most visible expressions of the industry are the huge power stations, mainly distributed around the coast (Fig. 1.2). These locations were chosen to efficiently draw on cold sea water for cooling, but this spread also reflected official nervousness surrounding the new technology and the requirement to locate stations in areas of low population density. An underlying factor in this national distribution was the availability of recently relinquished wartime airfields and munitions works, which provided large sites with basic infrastructure. This legacy will be discussed more fully in the individual chapters. In addition to the large power-generating reactors, the research reactors located in the south Midlands, south of England, and the outliers at Dounreay (Caithness) represented one of the most diverse groups of reactor types on the international stage. Reactors are only part of the story, and the national capability to manufacture nuclear weapons, power stations and research machines was underpinned by existing chemical and engineering companies, new firms that evolved to fill specialist niches, and others that exploited unexpected crossover technologies. A detailed appreciation of the nuclear supply chain remains to be investigated.

Politicians saw the taming of the atom as a means of maintaining Britain's global political status and improving her economic position, as well as providing the country with a reliable and clean energy source. It is over 70 years since work began on the pioneering Calder Hall (Cumbria) nuclear power station and its later

supply of electricity to the National Grid, followed by the Magnox reactor fleet and subsequent advanced cooled reactors (AGRs). At the outset many of these stations had an estimated economic life of 20 to 30 years. Although many exceeded this prediction, the Magnox stations are now decommissioned, and within a decade most of the later AGRs will also come to the end of their time.

As with any complex technological plant or machine, such as ships and aircraft, over a lifespan of many decades many improvements, including radical transplantation of major components and the introduction of new information handling systems, are to be expected. So too within nuclear power stations, where original tried and tested control systems sat alongside the most up-to-date monitoring equipment.

Fig. 1.1
Science Museum, London. Atomic-themed screen at the entrance to the Science Museum during the Festival of Britain 1951.
[© Historic England Archive, AA093673]

Fig. 1.2
Map of the principal nuclear facilities in the United Kingdom.
[© Historic England]

Other modifications might be made to the internal configuration of individual machines, such as installing new turbine blades, which had a marked impact on their performance but left few, if any, external traces. Improved scientific and engineering knowledge and safety legislation and regulations have led to further changes. Minor alterations might also be made to the structures. For example, in one of the Calder Hall reactors, a wall was thickened to give greater protection to an adjoining rest room. Such information is recoverable by what we recognise as standard methods of archaeological and architectural investigation, but was generally outside of the resources of this project. Beyond academic interest, this form of buildings archaeology is a crucial stage in nuclear decommissioning, where, especially in the early plants, built reality may differ from the blueprints.

Unlike in most industries, the activity of decommissioning and demolition in the nuclear industry often leads to new construction, including accommodation for contractors and secure storage facilities. Specialised robotic machines guided by electronically surveyed point clouds are also required to undertake the dismantling of highly contaminated plant. As part of the decommissioning regime, to allow natural radioactive decay of irradiated reactor components and other low-level waste, most reactor buildings will be sealed for a generation.

In the study of the recent material culture, oral testimony is increasingly being added to established archaeological methods of investigation. No oral evidence was systematically collected for this project, although the information given by members of staff who have acted as guides around various facilities has been invaluable. Within the industry some interviews have been conducted to capture the knowledge of past practices and processes but are generally not readily accessible to external researchers. The most comprehensive oral history project of the industry to date was undertaken in 2010 by the Whitehaven Archive and Local Study Centre (Cumbria), which interviewed around a hundred people associated with the Sellafield nuclear site.[6]

This book is about the built heritage of the nuclear industry. It is a contested heritage; to some these installations represent a heroic period of post-war industrial and technical prowess, while critics challenged the morality, cost and effectiveness of a defence policy constructed around nuclear weapons. This was an era when the country applied its scientific resources to develop new technologies in collaboration with the engineering industry, which allowed the boast that it was an industry 'built in Britain'. Civil applications of nuclear technology were often caught in this argument, while others challenged its safety, environmental impacts, economics and unwelcome intrusion into often undeveloped countryside.

Readers who wish to begin to explore the policies and economics of nuclear power in greater depth might begin with Simon Taylor's *The Fall and Rise of Nuclear Power in Britain: A History* (2016). Similarly, those who would like to know more about individual reactors should consult Nicholas Hill's overview *An Atomic Empire* (2013).

All measurements are in metric, with conversions from imperial values given to two decimal places. Every effort has been made to obtain high-quality images; unfortunately, for some historic pictures of now lost buildings and plant, only relatively low-resolution images were accessible.

2

Beginnings from laboratory to factory

Scientific origins

The current understanding of the atom is just over a century old. In the last decade of the 19th century, major intellectual advances were made across Europe that laid the foundations of modern atomic physics. One of the most significant discoveries – that of X-rays' ability to pass through solid matter – was demonstrated by Wilhelm Röntgen in Würzburg, Germany, in 1895. Around the same time, in Paris, Henri Becquerel observed the phenomenon of photographic plates being fogged by streams of particles produced by uranium, and a few years later, Pierre and Marie Curie announced that they had separated radium from pitchblende. Shortly after the turn of the century, in 1905, Albert Einstein famously published his special theory of relativity, $E=mc^2$, which proposed that enormous amounts of energy were stored within matter.

In Britain, during the late 19th and early 20th centuries, the Cavendish Laboratory, Cambridge, occupied a leading position in Europe's scientific circles and was one of the most influential centres for research into atomic physics. It was housed in a purpose-built laboratory, designed by W. M. Fawcett and opened in 1874, with extensions by the same architect in 1896 and 1908, and a further addition in 1936 (Fig. 2.1). Physicists working in this building were responsible for some of the most significant advances in atomic theory of their age. In 1897, Professor Joseph John (J. J.) Thomson demonstrated that cathode rays were negatively charged streams of particles, which he called 'electrons'. Two years earlier he had been joined by the New Zealander Ernest Rutherford, who would go on to transform the concept of the atom. Early ideas about the form of atoms imagined them as solid balls; indeed, by the beginning of the 20th century, an atom was commonly conceived as resembling a plum pudding, wherein the electrons were the plums. In 1911, after time in Canada, Rutherford, by then at Owens College, Manchester, revolutionised this model by proposing that an atom was not solid but comprised a nucleus surrounded by electrons. His move to this position illustrates the importance of world-class experimental facilities in attracting the leading researchers. In Manchester, this was represented by the new Physical Laboratory with its up-to-date experimental facilities designed by William Beaumont in collaboration with Professor Arthur Schuster.[1] It is listed Grade II. Rutherford's theory was further refined in 1913 by the Danish physicist Niels Bohr, who conceived of a simple model of a hydrogen atom with an electron orbiting its nucleus, a theory that remains useful today.[2]

Rutherford continued to be fascinated with the properties of the atom, and after his return to the Cavendish as its director in 1919, he single-mindedly pursued this theme. By the late 1920s, John Cockcroft, who had been briefly taught by Rutherford in Manchester, had also moved to the Cavendish and was directed by Rutherford to investigate the structure of the atom and if it might be possible to artificially disintegrate, or split it.[3] The final successful outcome was achieved through a combination of factors. Atomic theory was advancing rapidly at this time, and the work by the Russian Georgii Gamow pointed the way forward for Cockcroft. New laboratory equipment was also required, including a means of stepping electricity up to very high voltages. Here the laboratory was able to draw on work by the large electrical manufacturers then beginning to design equipment for long-distance power transmission. The transfer of technology from the large industrial concerns was aided by close personal links between the laboratory and the Metropolitan-Vickers Electrical Company, Stretford, Manchester, where Cockcroft had spent time as a college apprentice.[4]

The early experiments to split the atom in 1930 and 1931 were unsuccessful, and work was temporarily interrupted by a reorganisation of the laboratory. This allowed Cockcroft and his new assistant, Ernest Walton, in autumn 1931 to redesign their condenser/rectifier device and install it to the rear of the Cavendish in a far larger room previously used as a lecture theatre in the Balfour Building.[5] This new equipment cost around £1,000 and, at the time, was the most expensive item the laboratory had purchased.[6] In April 1932, using this apparatus, it was found that by bombarding a lithium target with artificially generated protons they were able to observe lithium, with a mass of seven, occasionally absorbing a proton and the resulting nucleus, with a mass of eight, breaking down into two helium nuclei. They had split the atom. Typical of experimental equipment, it was dismantled and dispersed once it has served its purpose. However, the significance of this work was recognised, and in the following year the Science Museum acquired some parts of the machine.[7] In later years the modified lecture room was used as a store whose fittings sadly removed any trace of past activities.

In this pioneering atmosphere, earlier in the same year, the existence of the neutron had also been confirmed by James Chadwick.[8] Success had been achieved through the advances in atomic theory, the availability of new laboratory equipment, and the work of the scientific community at the Cavendish led by the driving force of Rutherford. While at Cambridge Cockcroft also gained valuable practical experience of overseeing the construction of large scientific premises. In the early 1930s, he was closely involved with the design of the Royal Society Mond Laboratory in Cambridge, listed Grade II.[9] Later, in 1936, he oversaw the construction of the Austin Wing to the rear of the Cavendish, financed by a large donation from the car manufacturer Sir Herbert Austin (Fig. 2.2).[10]

The 1930s were marked by many important achievements for scientists working in Britain, but the physicists at the Cavendish and elsewhere were acutely aware that other groups might also be close to splitting the atom. Prior to the First World War, British, French, German and Russian scientists had led the field of atomic physics. In the immediate aftermath of the war, with the Continent in disarray, Britain dominated the subject; however, as the post-war decade advanced, this national lead was challenged by several European centres and increasingly by North American universities. As in numerous areas of scientific endeavour, many researchers regarded themselves as part of an international community. They reported on their latest discoveries in freely accessible scientific publications and exchanged knowledge through personal contacts, visits, symposia and conferences. In Berlin in late 1938, Otto Hahn and Fritz Strassmann observed that uranium bombarded with neutrons produced radioactive isotopes (*see* Box 2.1, p. 11).[11] They did not fully appreciate the significance of their discovery, but described it to their former colleague Lise Meitner, an Austrian, and now a refugee from

Fig. 2.1
The Cavendish Laboratory, Free School Lane, Cambridge. This was the first building in the world devoted to physical research and one of the early European centres for the investigation of nuclear physics. Listed Grade II. [© Historic England Archive, DP440452]

Fig. 2.2
High Tension Laboratory, Cavendish Laboratory, Cambridge, August 1938. 1 and 2MV Cockcroft-Walton accelerators, purchased from the Philips Company. These devices survived until the early 1960s when this space was converted into a lecture theatre and laboratories.
[© Historic England Archive, CC014174]

Nazi Germany living in Sweden. She in turn discussed the discovery with her nephew Otto Frisch, also a physicist and a fellow Austrian émigré, who coined the phrase 'fission' to describe the phenomenon. The experiment was reported in early 1939 and was quickly replicated in many other laboratories.[12] In April of that year, the Joilet-Curie team in Paris was the first to recognise that not only were fission fragments and energy produced during the experiment, but also enough neutrons to maintain the process, and quickly published their discovery. Later in the year, on 1 September, as German forces were poised to invade Poland, Niels Bohr and John Wheeler, from Princeton University, published a joint paper in *Physical Review*, noting that fission was far more likely to occur in uranium-235 (U-235) atoms rather than the more common uranium-238 (U-238) atoms.

Box 2.1 Some basic physics

An atom is made up of protons, neutrons and electrons (Fig. B2.1.1). The proton has a positive electrical charge, the electron a negative charge, and, as its name suggests, the neutron is neutral. Protons and neutrons form the core of the nucleus of an atom around which the electrons rotate in various orbits. Together, neutrons and protons are known as nucleons, the neutrons having the effect of aiding the positively charged protons to bind together. Normally an atom will have an equal number of protons and electrons, making it electrically neutral. The mass of an atom is calculated by adding the number of protons and neutrons together; thus helium (*see* Fig. B2.1.1) has a mass of four. The atomic number of an element is defined by the number of its protons; hence, helium has an atomic number of two. Most species of atoms may be characterised by their atomic and mass number, or by their name and mass number: this is referred to as a nuclide. Helium-4 is a nuclide comprising two protons and two neutrons. In a stable nucleus, the number of protons and neutrons will be roughly equal. Unstable nuclei, with an excess of neutrons, tend to transform themselves to a more stable form by the emission of a negatively charged electron, a beta particle, thereby converting a neutron into a proton. Where there is an excess of protons, by losing positive charge through the emission of a positron they are converted into a neutron. Any excess energy left in the nucleus is lost through the emission of gamma rays, discrete quantities of energy without mass or charge. These processes are known as radioactivity.[1]

In any given chemical element, the number of protons in the nucleus is constant, but the number of neutrons may vary, resulting in a different mass. An element may therefore have several types of atoms (Fig. B2.1.2). These are known as isotopes; some occur naturally, while others are artificially created in a reactor by bombarding atoms with neutrons. Isotopes are referred to by the number of the nucleons; uranium-238 (U-238), for example, contains 92 protons and 146 neutrons.

The principles of nuclear power

Thermal nuclear reactors generate heat that is converted into pressurised steam to power turbines, which, in turn, generate electricity. The heat is produced by the continuous splitting of atoms, a process known as 'fission' (Fig. B2.1.3). Some types of heavy nuclei, uranium and thorium, experience spontaneous fission; others may be made to fission by the addition of energy, for example by neutron bombardment. The only material that is naturally fissile to a significant extent is uranium-235 (U-235), which comprises just 0.7 per cent of natural uranium.

During the fission process, energy is released,

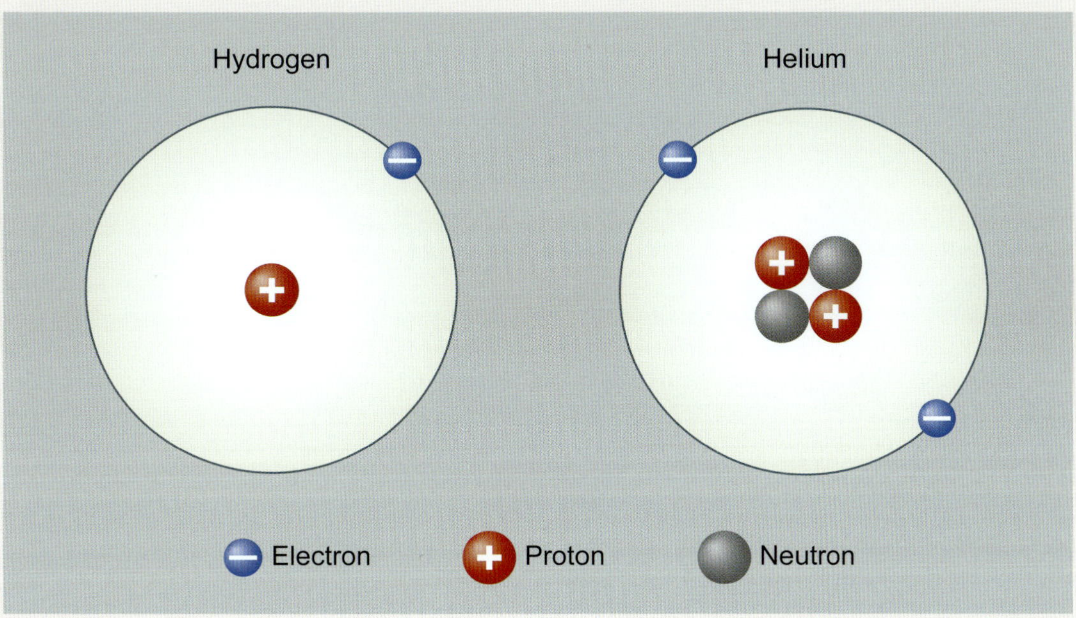

Fig. B2.1.1
Atomic structure.
[© Historic England]

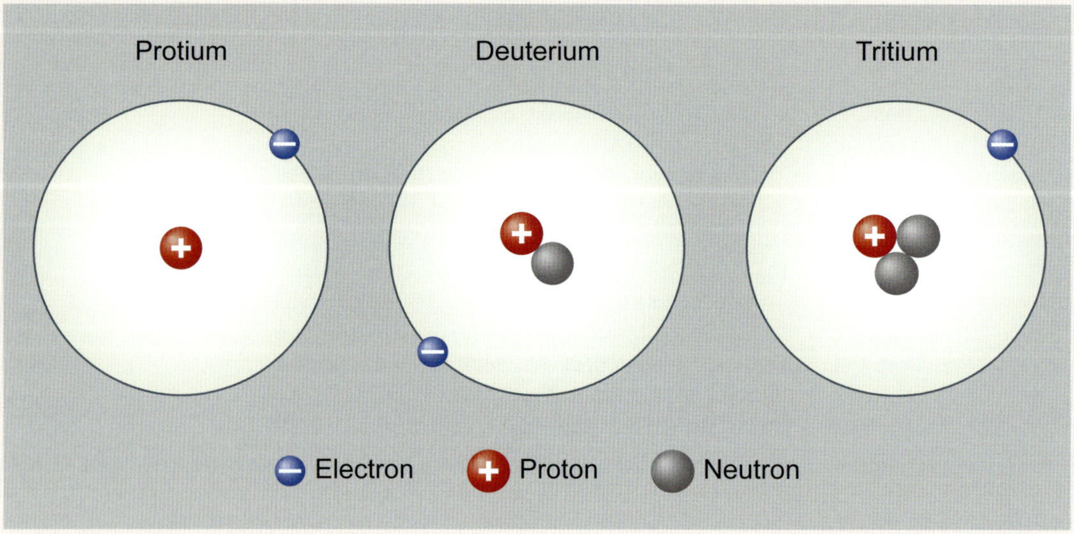

Fig. B2.1.2
Nuclear isotopes.
[© Historic England]

mainly in the form of kinetic energy (the energy of motion) of the fission fragments. This is rapidly converted to thermal energy, which causes the temperature of the fuel to rise. This heat may then be extracted by a coolant, and may be used to generate steam through a heat transfer system.

During the fission process, the unstable fission fragments give off neutrons. These may be split into two types: prompt neutrons, which are produced almost instantaneously, and delayed neutrons, that may be emitted seconds or minutes later. The emission of neutrons has several consequences. First, it allows a nuclear chain reaction to be sustained. Second, it results in the creation of transuranic elements (those with an atomic number greater than uranium), produced in non-fission reactions by the absorption of neutrons by the uranium fuel. It is by this method and beta decay that plutonium is produced as a by-product of power generation. Neutron absorption by U-238 produces U-239, which produces neptunium-239 and then plutonium-239 through beta decay. Third, neutrons cause the materials of the power plant to become radioactive.[2]

As we have seen above, in a nuclear chain reaction, fission results in the release of neutrons. Some, however, will be absorbed in non-fission reactions and others will be lost to the system. To minimise the loss of neutrons, the reactor core is constructed from materials that are less prone to absorbing neutrons and by lining the core with a reflector. A core also needs

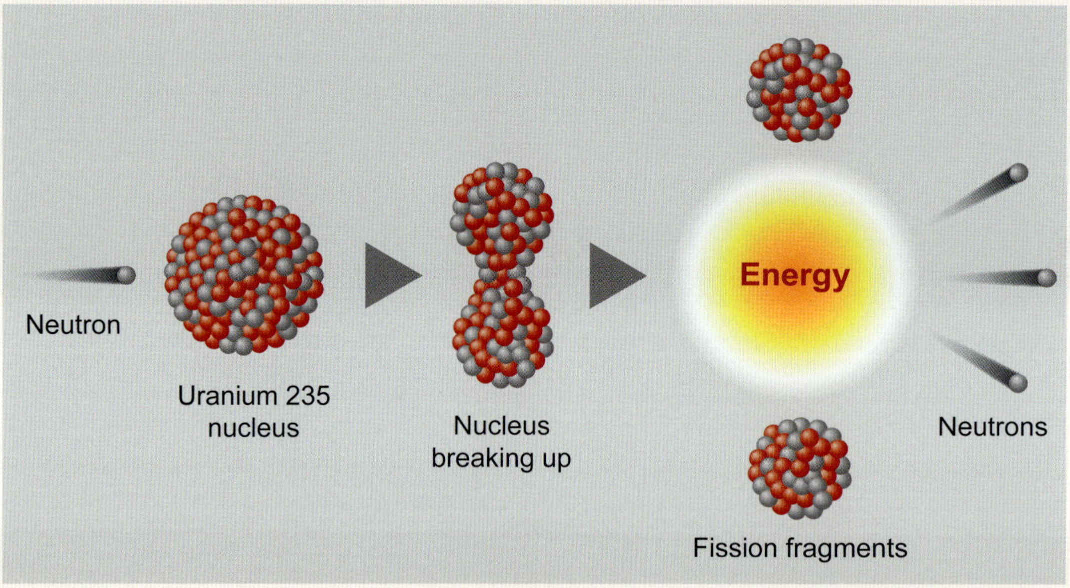

Fig. B2.1.3
Nuclear fission.
[© Historic England]

to reach a certain size, or critical mass, before a chain reaction may be sustained. In a reaction, if 2.3 neutrons are released and 1.3 are reabsorbed, 1 neutron remains to sustain a constant critical reaction. If the absorption rate were to rise to 1.31, then only 0.99 neutrons would be available for fission and the fission rate would decrease in this subcritical reaction. Conversely, if the absorption rate declined to 1.29, the fission rate would rise in supercritical reaction.[3]

Reactor types

The choice of reactor type used to generate power is the result of the dynamic interplay of technological, political, economic and cultural factors. The chief engineering considerations in the design of the reactor are its fuel, moderator and coolant. The most commonly used fuels are U-233, U-235 and plutonium; moderators may be of carbon (graphite), light hydrogen (ordinary water) and heavy hydrogen (heavy water); and for coolants, water, heavy water, liquid metals and gases (such as carbon dioxide and helium) are used.[4] Heavy water is a compound of oxygen and the heavy isotope of hydrogen, deuterium, and has a greater density than normal light water.

To control the chain reaction, a moderator is required to slow down the neutron bombardment and to regulate the amount of heat produced. Control rods made of boron or silver may be inserted between the fuel elements in the reactor core to absorb neutrons and control the chain reaction.[5] The heat generated during fission is removed by the coolant, which may be water, gas or liquid metals. This is then generally taken to heat exchangers where it is converted to steam to power the electricity-generating turbines. In some types of reactors, the water is allowed to boil in the core and the steam is passed directly through the turbines.

Reactors may be broken down into two main types: thermal reactors and fast reactors. In a reactor, the chain reaction, or fission, is maintained by the collision of neutrons with the uranium fuel. Neutrons generated by fission are referred to as ‘fast’ neutrons, and once they have been slowed down by the moderator they are known as ‘thermal’ neutrons. To make the collisions more likely, the neutrons are slowed down, or thermalised, by a moderator, often graphite or heavy water. Thermal reactors may therefore be further subdivided into two broad groups: graphite-moderated and water-moderated. In a fast reactor, there is no moderator.

The Second World War – academic beginnings

By the summer of 1939 a small group of British academics, bound together by past associations, and often aided by colleagues fleeing persecution on the Continent, had independently begun to explore various aspects of atomic phenomena, including its military application. At the Imperial College of Science and Technology (London), George Thomson, son of J. J. Thomson, began to investigate whether a continuous chain reaction could be produced using uranium oxide and a moderator of water or paraffin. The answer was negative, giving some comfort for a time that an atomic weapon was not imminent.[13]

James Chadwick moved to the University of Liverpool in 1935, and there, by the end of the decade, he was leading the construction of the country's first cyclotron, an accelerator where electrically charged particles are propelled in spiral paths by use of a magnetic field (the device is now on display at National Museums Liverpool).[14] He had been intrigued by the Bohr and Wheeler paper, and in the autumn of 1939 he had concluded that the fission of uranium by fast neutrons – that is, neutrons not slowed down by collisions or a moderator – might be a possibility. Towards the end of the year, conscious that he might not be alone in this proposition, he wrote to Professor E. V. Appleton, secretary of the Department of Scientific and Industrial Research, to enquire if anyone else was investigating fission, and contacted Thomson, who reported the negative results of his experiments. Under Chadwick's leadership important research was carried out on basic nuclear problems, and some of the earliest studies of fast neutrons that would make a bomb possible were conducted.[15]

Separately, Rudolph Peierls, a Berliner by birth, and recently appointed professor of Mathematical Physics at the University of Birmingham, began to calculate the amount of pure uranium that would be required to produce a critical mass (a self-sustaining chain reaction, *see* p. 13), and if it would lead to a violent release of energy. In the summer of 1939 he was joined by Otto Frisch.[16] The conclusions of Peierls' calculations were that, although it would be technically difficult and expensive, it was theoretically feasible to create a uranium bomb. To communicate this information to the British government on the advice of Marcus Oliphant, an Australian physicist working at Birmingham, Peierls and Frisch wrote to Sir Henry Tizard, rector of Imperial College and a member of the Air Ministry's Aeronautical Research Committee.[17] He in turn shared the memorandum, covering three sheets of foolscap and titled, 'On the Construction of a Super Bomb', with his colleague George Thomson.[18] In it, Peierls and Frisch argued that 5kg of U-235 would have the equivalent explosive power of thousands of tons of dynamite, in addition to the deadly effects of radiation. They also proposed a method whereby lighter and more fissionable U-235 might be separated from heavier U-238. Through the Air Ministry Committee, a small group chaired by George Thomson was established to investigate the feasibility of developing an atomic bomb during the present war. After initial scepticism in the scientific community, the so-called 'MAUD Committee' under Thomson reported in July 1941 confirming that it might be possible to manufacture a bomb using U-235; it also discussed the peaceful uses of atomic energy to produce power, a lower priority at this time.[19] In September, the prime minister, Winston Churchill, advised by the Chiefs of Staff, approved the decision to proceed with the development of an atomic bomb, perhaps with little conception of the scale of the undertaking required.

The British preference was for a uranium bomb, which would require devising a method to separate U-235 from heavier U-238 (*see* Box 2.1). This problem intrigued another German émigré, Professor Franz Simon at the Clarendon Physics Laboratory, Oxford; Simon was perhaps stimulated by conversations with Peierls on his visits to Oxford during the first winter of the war. In the spring of 1940, on the initiative of Simon and his colleagues, work began on the development of gaseous diffusion technology to separate isotopes using a fine copper gauze known as Dutch cloth. By the summer, the Oxford group had joined the embryonic network of scientists and engineers developing this new technology.[20] Peierls continued to lead a theoretical group at Birmingham. His colleague, Professor W. N. Haworth, later collaborated with ICI to manufacture the gas uranium hexafluoride, needed for the gas diffusion process, and the production of pure uranium metal.[21] Frisch subsequently left Birmingham for Liverpool, where in early 1941, assisted by Peierls and M. H. L. Pryce, he calculated the amount of material required to produce a critical mass vital to successful nuclear fission.[22]

Heavy water

After the fall of France to the Germans in summer 1940, Britain's lead in this field was strengthened by the arrival of two French scientists from the Joilet-Curie team in Paris, Hans von Halban and Lew Kowarski. The focus of their work had been on the development of a nuclear power reactor, or 'boiler', and, significantly, they brought with them the entire European stock of heavy water, weighing 185kg.[23] The importance of heavy water lies in its ability to be used as a moderator in an atomic pile, where it has the effect of slowing the passage of neutrons, thereby making a collision and fission more likely. During the Phoney War of early 1940, this precious cargo had been acquired by the French from the Norsk hydro plant at Rjukan, Telemark, Norway, and to avoid interception by the Germans was flown to Edinburgh and then taken by train to London. From here it travelled to France, a country still trusting in the strength of the Maginot Line fortifications to keep it secure. But by June this position was untenable, and after further frequent moves the French government ordered that it should be moved to England. Once ashore it was transported to the Cavendish Laboratory, Cambridge, where the French team worked before moving to Montreal and then to Chalk River (Ontario) (*see* p. 21).[24]

Ironically, many of the physicists whose discoveries gave Great Britain a conceptual advantage in the development of atomic weapons during the war were European émigrés, whose origins barred them from working on top-secret radar projects but left them free to explore atomic physics. The next challenge was to translate laboratory phenomena into industrial-scale processes.

Scaling up for industrial production

One of the few companies with the breadth of capabilities and resources to embark on a scheme as complex as the separation of uranium isotopes was Imperial Chemical Industries (ICI).[25] By the outbreak of war, through a series of mergers dating back to the 1920s, ICI had become a large conglomerate with chemical and related factories spread across the country. The professional knowledge of its staff and their practical ability to develop novel industrial processes would be crucial to the separation of uranium isotopes on an industrial scale.

ICI had been involved in initial discussions with the government in early summer 1940 about participating in atomic research. It was very keen to develop both separation and, potentially, reactor technology, partly through a sense of national duty, but also with an eye on the future commercial opportunities that atomic energy might offer. By this time the company was already undertaking diffusion research at its central chemical laboratory at Widnes (Cheshire).[26] Later, in October 1941, the Scientific Advisory Committee Defence Services Panel rebuffed ICI's desire to develop an atomic 'boiler', informing them that the government wished to retain control of all uranium work through the Department of Scientific and Industrial Research. In July of the same year, Sir John Anderson, Lord President of the Council and Chancellor of the Exchequer in Churchill's government, became the minister in charge of the atomic bomb project, providing the link between the research effort and the Cabinet.[27] Under the cover name of the Directorate of Tube Alloys, the work was led by Wallace Akers, formerly research director at ICI. Its headquarters was placed close to the centre of government, just off St James's Park, at 16 Old Queen Street, less than 200m from the Cabinet War Rooms (Fig. 2.3).[28]

Fig. 2.3
16 Old Queen Street, London, a few streets away from Whitehall. This building was the London headquarters for the Tube Alloys project.
[© Historic England Archive, DP462647]

The diffusion challenge

One of the most urgent tasks was to devise a process to separate the lighter fissionable U-235, comprising about 0.7 per cent of natural uranium, from the heavier U-238. Potentially there were several methods that might be used, but, on the recommendation of Professor Simon, in May 1941 it was agreed to build a gaseous diffusion plant.[29] This process relied on the lighter molecules of U-235 diffusing more quickly than the heavier U-238 molecules through a porous membrane, with millions of very fine holes. If the gas that had diffused through the membrane was then pumped away, it would contain a greater number of lighter molecules. For this procedure a gaseous form of uranium was required, which was provided by uranium hexafluoride, or hex.

Building on the Dutch cloth method, an early challenge was to produce a porous membrane which an initial estimate calculated might require 160,000 holes per square inch.[30] Dr Arms, a colleague of Professor Simon at the Clarendon Laboratory, approached ICI for assistance and was directed to their research manager S. S. Smith in the Metals Division at Witton, Birmingham. Initially, mechanical methods were explored to produce a fine mesh by beating bronze gauze, even harnessing the skills of an elderly gold beater, presumably from the local jewellery industry, but without success. Smith, inspired by seeing the dots of a half-tone photograph blown up on a poster, approached a colleague, Michael Clapham, at ICI's Kynoch Press to discuss if a printing and photographic etching technique might provide a solution to the problem of manufacturing such a fine membrane. A negative was produced but plans to print it locally were thwarted by the bombing of the intended print works, and the job was moved to one of the country's leading printers, the Sun Engraving Company, in their now demolished works in Whippendell Road, Watford (Hertfordshire). Some progress was made with this method, and throughout 1942 prototype membranes were produced. In the search for a more uniform product with accurate hole sizes, Clapham turned to his former employer, the printers Lund Humphries of Priestman Street, Bradford, where a derelict stone façade of their premises remains. This company used the Aller offset-litho printing process, which applied an extremely thin copper film to stainless steel plates by electro-deposition. This was an important step forward, but one ultimately succeeded by powder metallurgy processes.[31]

Ministry of Supply Valley, Rhydymwyn, Flintshire

Beyond original archival documents, the early work on the atomic bomb in Britain has left few tangible traces. Uniquely, at Rhydymwyn some evidence remains of the efforts to transform gaseous diffusion from a scientific theory and laboratory bench experiments into an industrial process. Uranium hexafluoride, or hex, which had been identified as the best gaseous medium, is an extremely hazardous substance that reacts violently with water, is toxic, and corrosive to most metals. Machines to handle and compress this gas had to be highly resistant to corrosion and leak-proof.[32] Initially one two-stage unit and two ten-stage units were ordered from the Metropolitan-Vickers Electrical Company, Manchester. Surviving drawings indicate that design work on the machines began in late 1941.[33] However, in early 1942, to address worries about the configuration of the machines, two single-stage trial units were added to the contract, to refine the development of the more complex units. At first it was proposed to site these units at the Clarendon Laboratory, but subsequently an ICI team at Billingham, Stockton-on-Tees, led the development work, and space was identified within a Ministry of Supply (MoS) mustard gas factory operated on the government's behalf by ICI at Rhydymwyn. Safely tucked away in rural Wales 5km from Mold, 'MoS Valley', as it was termed, was less threatened by air raids. Also, an adjacent railway station provided convenient access for academic experts from the universities of Oxford, Birmingham and Liverpool, and their industrial collaborators from ICI and Metropolitan-Vickers located in the north-west of England. The factory was staffed by about 20 scientific workers from the Clarendon Laboratory, a similar number from ICI Billingham, and around 60–70 labourers.[34]

The construction of MoS Valley had begun just prior to the outbreak of war, but by early 1941 it was recognised that production capacity of mustard gas far exceeded foreseeable demand (Fig. 2.4). One consequence was that a large production building, P6, measuring 64m by 27.5m, remained vacant (Fig. 2.5). This was ideal for secret experimental work, offering high security and a large open floor area for setting out machinery. Anonymity was also maintained by use of the codename Project X.[35] P6 survives today. It comprises seven bays, the roofs of the central five bays formed of curved reinforced concrete, with flat roofs over the end bays. Above the southern end is a concrete tower that perhaps held a water tank. Longitudinally, the building is divided into five uneven bays and has a central brick wall.[36] Overhead gantry cranes enabled the installation of plant and movement of materials.

Before work could begin, some modifications were required to this building. Vents and doorways were sealed, and a more powerful heating system was installed. Along the east side offices

Fig. 2.4
Rhydymwyn, Flintshire. View showing the three large mustard gas production buildings; P6 is to the top right. The former railway line is marked by the fence line to the rear of the modern industrial estate.
[© Crown copyright. RCAHMW, 6185494]

Fig. 2.5
Rhydymwyn, Flintshire, Ministry of Supply factory building P6, July 1997. The cleaner-looking bricks represent post-war blocking of wartime doors and windows; traces of camouflage may still be seen.
[© P. W. Bone]

and laboratories were created with windows inserted into the external wall. Access doors, presumably for the installation of machinery and the movement of chemicals, were also introduced. Evidence from scars on the floor suggests that the experiments were concentrated in the eastern side of the building, adjacent to offices and laboratories.[37]

In December 1941, Metropolitan-Vickers estimated that a two-stage machine to test its main components could be operational at Rhydymwyn by August 1942, and the ten-stage machines by the end of the year. In a reflection of the complex and novel engineering involved, the single-stage machine, ordered as an afterthought, was first run on New Year's Day 1943, and the two-stage machines in the following August. The ten-stage machines were delivered in November, installed and tested, but by this date the Quebec Agreement (*see* p. 20) had been signed and the focus of research activity had moved to North America. Some investigations probably continued at Rhydymwyn and resumed fully after the war, when there are indications that some of the machines were relocated to the Atomic Energy Research Establishment,

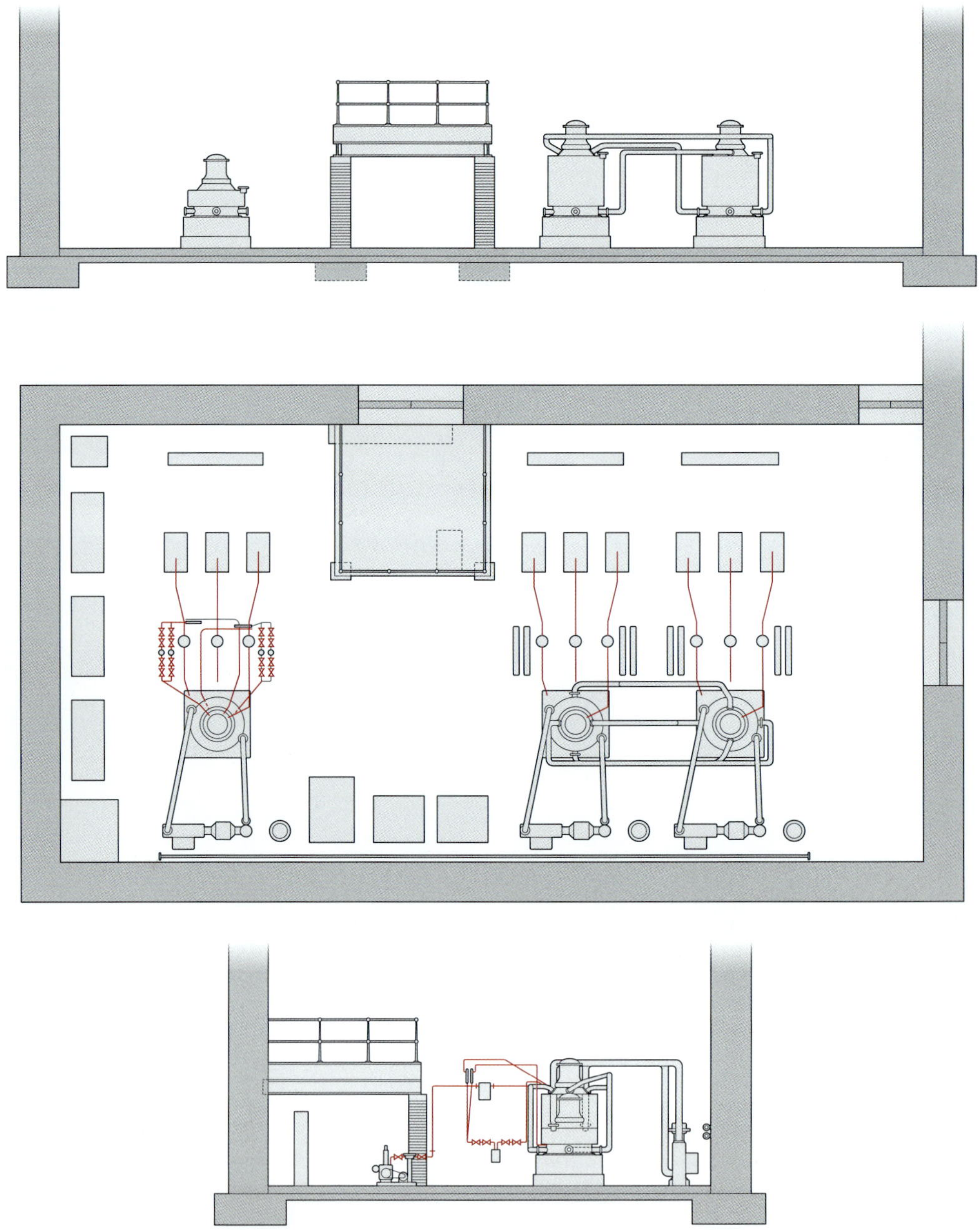

Fig. 2.6
Rhydymwyn, Flintshire. Design drawings for gaseous diffusion machines produced by Metropolitan-Vickers. No scale was given on the drawing. Each machine was about 1.2m in diameter.
[Redrawn from TNA AB1/68]

Harwell (Oxfordshire), and later to another former ICI agency factory at Springfields (Lancashire) (*see* p. 25).[38]

In total, two single-stage, one two-stage, and two ten-stage machines were installed at Rhydymwyn. Drawings produced by Metropolitan-Vickers, which survive in The National Archives, show the general arrangement of the machines, although it is unclear if these relate exactly to the pattern of installations in P6 (Fig. 2.6). It has been estimated that the machines were about 1.2m in diameter at the base and the internal impellor was around 72cm in diameter. This comprised a rotor driven at high speed by a motor which created a centrifugal force which could be used to separate the heavier uranium isotopes. Each machine weighed about 3 tonnes. The immediate requirement for the work at Rhydymwyn was for the electrical and mechanical testing of the equipment and proof testing of the porous membranes. Some hydrodynamic trials with non-toxic and non-flammable sulphur hexafluoride may have taken place, but the absence of evidence for an associated chemical plant at this early stage suggests experiments with hazardous uranium hexafluoride were not yet contemplated.[39]

The trials plant at Valley, in particular the alterations carried out to the fabric of P6 and the scars of machinery bases on its floor and walls, represents some of the earliest physical traces of the nuclear industry to be found anywhere in the world. The building is listed Grade II, and occasionally open for guided tours. However, to understand the enormity of the task ahead and the cost in resources to develop this technology further, it is necessary to look across the Atlantic to the near-contemporary American K-25 gaseous diffusion plant at the Clinton Engineer Works, Oak Ridge (Tennessee). Here, a huge U-shaped structure, over 0.8km in length and 914m wide, was constructed between 1943 and 1945. It was then the largest building in the world, with a workforce of over 10,000 people. The plant operated until the late 1980s and was finally demolished in 2013.[40] If such a facility had been built in Britain, it would have quickly attracted the attention of the German air force.

Despite wartime demands, ICI's Metals and General Chemicals divisions continued to pursue an interest in developing the peacetime application of the technology – a nuclear 'boiler'. In Widnes (Cheshire) by mid-1943, trials had advanced to the stage where they were able to send uranium billets weighing 90.72kg (200lb) to their Metals Division at Witton, where further experimentation was undertaken to produce uranium fuel rods. At Billingham, investigations also began into heavy water production, which might act as a moderator in a reactor.[41] After August 1943, as the focus of British atomic bomb work moved to North America, and with it personnel and resources (*see* following subsection), this activity tailed off.

By June 1944 the government had authorised ICI to spend £975,000 on 'Tube Alloys', of which £870,000 had been disbursed; the greatest proportion of the expenditure was spent at Billingham, where a team had led early research on gaseous diffusion.[42] This sum represented nearly two-thirds of the government-funded wartime research undertaken by ICI to that date, a company that through its production of chemicals and explosives was crucial to the war effort. It was, however, insignificant compared to the $2.2 billion expended by the United States on its atomic bomb project. In comparison this was more than four times the amount the Germans spent on the V2 rocket programme.[43] Many aspects of ICI atomic research during the war are obscure. Their official historian, W. J. Reader, noted the 'paper records were either destroyed or swallowed by official secrecy'.[44]

The American Manhattan Project

At the same time as the European scientists were exploring nuclear fission, several scientists in the United States were also considering the theoretical possibility of constructing an atomic bomb. As in Britain, many of their leading figures were European refugees, who were aware of nuclear research in Germany and of the dangers it might pose. Leo Szilard, a Hungarian émigré, was so concerned that in 1939 he wrote a letter to President Roosevelt alerting him to the perils, and to give it more credibility he arranged for it to be signed by Albert Einstein.

As the war advanced there was initial scepticism on both sides of the Atlantic that an atomic bomb could be constructed in time to be of use, and of a size that could be delivered by an aircraft. In summer 1940, an influential British scientific mission, headed by Henry Tizard, visited America to discuss recent developments in military technology. Nuclear fission and the world's uranium stocks formed only a very small part of these talks, reflecting the doubts in the minds of leading scientists, including John Cockcroft, that a viable weapon was achievable in a few years. One outcome of these exchanges was the establishment of an office of the United States National Defense Research Committee (NDRC) in London. A member of this mission, Kenneth Bainbridge, was given special permission to attend meetings of the top-secret MAUD Committee. Through this contact he became conscious of a distinct shift in British scientific opinion that a bomb might be feasible. Reports reaching his superior, Vannevar Bush, then chairman of the NDRC, together with informal contacts between British and American scientists, and the publication of the final MAUD report, helped to convince Bush that the United States government should expand its nuclear research programme. The MAUD report was officially handed over to the United States in October 1941, and shortly afterwards George Pegram and Harold Urey were sent to England to learn about the work being carried out on uranium.[45]

In September 1942, Major General Leslie Groves, fresh from managing the construction of the Pentagon, Washington DC, was placed in charge of the American atomic bomb project, known as the 'Manhattan Project', after its initial New York office, with its headquarters at the remote township of Los Alamos (New Mexico) (Fig. 2.7). The physicist Robert Oppenheimer, University of California, Berkeley, led the scientific efforts. As the American programme gathered pace, British scientists found themselves increasingly excluded from key areas.[46] The Quebec Agreement signed between Britain

Fig. 2.7
Los Alamos, New Mexico. Larger-than-life statues of Robert Oppenheimer (left) and General Leslie Groves. [© Wayne Cocroft]

and the United States on 19 August 1943 allowed British personnel some access to the project, on an understanding that the bomb would not be used against one another, or a third party, or information disclosed without mutual consent.[47] It was from then that many of Britain's specialists working in this field, including many recently naturalised as British subjects, were transferred to the United States.

The Canadian connection

Earlier, during the Tizard mission, Cockcroft visited Canada on several occasions, principally in connection with his then responsibilities to develop and manufacture radar equipment, although on one trip in November 1941 he was also able to evaluate the progress of Canadian nuclear physics and was particularly impressed with the work of G. C. Laurence and E. W. Stedman. These contacts and with key officials gave the British the confidence in September 1942 to propose that the British and French nuclear research teams focused on building a heavy water-moderated uranium reactor should move to Canada.[48] Dr C. J. Mackenzie, president of the Canadian National Research Council, welcomed this proposal and quickly found accommodation for the team at the University of Montreal. Canada offered safety from the war and was also an important source of uranium ore. On 13 April 1944 it was agreed by the joint American and British policy committee that a heavy water-moderated pile should be built in Canada. Meanwhile, at the end of 1943, Cockcroft had been approached by James Chadwick, his former Cambridge colleague and now head of the British mission to the Manhattan Project, to ask him if he would accept the post of director of the Montreal team, a proposal he accepted. Cockcroft arrived in April 1944 and a site for the reactor was chosen at Chalk River on the Ottawa River about 180km north-west of Ottawa. In Montreal, design work began on a 10MW heavy water-moderated reactor NRX and ZEEP (Zero Energy Experimental Pile) while Cockcroft supervised the building of the new research establishment and settlement of Deep River. American support for this work only went so far. Detailed information on the structure of the Hanford (Washington) plutonium production pile and methods for extracting plutonium, plutonium chemistry and purification were withheld. One concession was the provision of a small number of irradiated uranium slugs from the American reactor at the Clinton Engineer Works (Tennessee) for the Chalk River team to experiment with. Research also began to develop chemical processes to recover uranium from irradiated fuel and the separation of U-233.

Conclusion

Wartime experiences in Canada had provided British, Commonwealth and French scientists and engineers with direct practical knowledge of reactor construction, sourcing materials and chemical extraction methods. This understanding became especially important after the passing of the American Atomic Energy Act (McMahon Act) in 1946, which effectively ended the exchange of nuclear information between the wartime allies. In post-war Britain, the knowledge transferred from Canada saved many months of work, as well as laying the foundations for Canadian and French post-war nuclear programmes.[49] The nascent British wartime efforts to develop an atomic bomb had been focused on producing a gun-type weapon using enriched uranium. In such a device were two subcritical masses termed the projectile and target. To create a supercritical mass and an explosive nuclear chain reaction, the projectile was shot at the target at high velocity using a conventional explosive propellant.[50] It was such a weapon, 'Little Boy', that devastated Hiroshima on 6 August 1945. As early as 1940, Britain had recognised the feasibility of a plutonium bomb, but the resources to develop technologies to produce enough fissile material and a detonation system then lay beyond reach.[51] The British team attached to the Manhattan Project subsequently learnt, and the detonation of 'Fat Man' over Nagasaki on 9 August 1945 clearly demonstrated, that a plutonium weapon could create a far more powerful explosion using less fissile material. The next chapter will explore the scientific and industrial facilities required to create such a device.

3

Cold War imperatives

Introduction

This chapter will describe the facilities that were established in the immediate post-war years to develop, manufacture and test the first British atomic and hydrogen bombs. Fundamental and applied research, and prototype development work, were carried out in defence research establishments chiefly located in the south of England. Alongside this, mainly in the North West, a new industry grew to provide fissile material. Initially, the principal establishments involved in the design of the bomb were the Armament Research Department, Woolwich, which oversaw the development of the bomb's high explosives and electronics components, and the Royal Aircraft Establishment, Farnborough (Hampshire), which was responsible for its ballistic casing.[1] The Atomic Energy Research Establishment (AERE), Harwell (Oxfordshire), contributed crucial input into the project, especially in supporting the Production Group in the development of processes and plant. The country's existing chemical and engineering firms provided vital industrial capability to supply specialist plant, materials, components and personnel.

Soon after the end of the war in Europe, discussions at the heart of government edged towards the decision to build an atomic bomb. Central to many of these discussions, and with links to the wartime Tube Alloys project, was Sir John Anderson, Chancellor of the Exchequer, who in June 1945 recommended the founding of a research establishment to investigate all aspects of atomic energy. Notwithstanding Anderson's loss of office in the July general election, the incoming Labour prime minister, Clement Attlee, appointed him to the new Advisory Committee on Atomic Energy. In October 1945 a Cabinet paper recommended the production of atomic weapons. A month or so later Anderson proposed an immediate start on two production piles, with the implication that a plutonium weapon was envisaged. The formal decision to proceed with an atomic bomb was taken on 8 January 1947 by the small, anonymously named, Cabinet Committee GEN 163.[2]

Industrial production

In parallel with the scientific research into developing the bomb's explosive and electronic components, and supporting theoretical work, which has left few physical traces, the manufacture of the fissile material at the heart of the bomb project required the creation of large and innovative production facilities. Christopher Hinton, an engineer who had previously worked for Imperial Chemical Industries (ICI) and had overseen the wartime Royal Ordnance Filling Factories, was placed in charge of the new Production Division. He was joined by William Leonard Owen, a former wartime colleague, as his deputy. The question then arose of where this new industry was to be located. It would require access to a skilled technical workforce and to engineering companies capable of constructing novel plant, often in competition with the private sector keen to recover from wartime controls and re-establish the export trade. Mindful of wartime bombing, locations in the west of Britain were preferred. Probably influenced by Hinton's earlier associations with ICI and his recent wartime experiences, the former filling factory at Risley (Warrington Unitary Authority) was chosen as the headquarters of the Production Division. The initial staff contingent, comprising 12 men and women, reported for work on 4 February 1946.[3] From these beginnings it became the centre for the design, construction and operation of the production factories, including all their buildings and plant.[4] Although, it was nearly another year before the formal decision to build an atomic bomb was taken, the new organisation was tasked with laying the foundations: building a uranium production plant, plutonium production piles and a chemical separation plant.[5]

Work was begun with a wartime urgency to non-negotiable deadlines. From the small advance party, teams of engineers, architects and designers were assembled and site preparation began, often before the exact plant specifications were known. The sequence of factory design followed the production flow from uranium refining to the commissioning of the first atomic pile.

In addition to research undertaken at AERE Harwell, the Industrial Group had its own dedicated research and development branch undertaking applied research for industrial needs comprising about 800 staff, with facilities at each of the Group's establishments except for Risley. Here, to support the Engineering Design Branch, a research section was accommodated in the temporary buildings of HMS *Ariel*, a wartime training establishment for naval radio and radar technicians about 2km to the north of Risley in Wigshaw Lane, Culcheth (Warrington Unitary Authority). Research was focused on reactor materials and the challenges of operating at very high temperatures and pressures and their effect on reactor constituents, including coolants, moderators, fuel and construction components, in addition to advancing the understanding of heat transfer and fluid flows. Important research was also conducted into the production of metals and alloys and the dimensional changes experienced by irradiated uranium and its effect on its cannister.[6]

Springfields Works

The first stage of the production process was to extract uranium from its ore, which was at this time acquired from the Belgian Congo (now the Democratic Republic of Congo). Prior to its use for atomic energy, its only commercial use had been as a colourant in glass and ceramics. The location chosen for this activity was a former poison gas factory built and operated by ICI on behalf of the government at Springfields, Salwick (Lancashire) (Fig. 3.1). This works was only a few years old and offered a basic infrastructure, including office accommodation, mains water, electricity and a rail connection.

Fig. 3.1
Springfields, Lancashire, 5 June 1950. Former wartime mustard gas factory converted to manufacture nuclear fuel. [Historic England Archive (RAF Photography), raf_541_561_rp_3073]

The refining of uranium ore was essentially a chemical engineering process whereby finely ground ore was dissolved in acids and processed to remove any impurities that might later affect the efficiency of the reactor. Typically, for the development of new chemical processes, laboratory apparatus would be scaled up to a pilot plant before the final production plant was constructed. Due to the demanding deadlines, this stage was omitted, although fortunately the designers were able to draw on ICI's wartime experience at Widnes (Cheshire). The chief chemical design engineer for the factory was Charles Turner.[7]

The extraction of crude uranium oxide was undertaken in a single building using standard chemical engineering technology. In comparison with other industries, it was distinguished by its cleanliness and ventilation to extract uranium dust and radon gas that might be released

Fig. 3.2
Calder Hall, Springfields, Lancashire. Uranium rod manufacture: a rod being removed from the beta-quench apparatus after heat treatment to reduce the grain size of the metal. [Courtesy of the Nuclear Decommissioning Authority, AJ1464]

during processing. All workers entering the production areas were required to change into white overalls and don white caps, white rubber boots and gauntlets. Just a year after the first team had arrived at Risley, and against the backdrop of one of the harshest winters on record, in the middle of February 1947 the pulverising process, using ball mills, was turned over to the production team for testing, and the remainder of the buildings by May.

Ore was delivered to the factory in sealed metal containers, and after initial crushing was taken to a ball mill where it was pulverised in a steel drum containing steel balls. This is a standard industrial process and was used until recently in many power stations to pulverise coal. While in the ball mill water was added to create a slurry, which was then run into stainless steel acid-filled dissolvers, 2.44m high and 1.83m in diameter, each holding 4,546 litres. Through the addition of other chemicals, unwanted metals and radium were extracted and packed for reprocessing elsewhere. Further along in the process large quantities of hydrogen peroxide were added to facilitate the precipitation of crude uranium oxide. Inspired by wartime German propulsion experiments, increased demand for this hazardous substance was also being made by contemporary research into rocket and submarine propellants. Through further chemical processes the uranium oxide was purified and transformed into ammonium diuranate, which was then reduced in a furnace to create metallic uranium before being cast into billets.

The resulting billets were either stored or moved to the casting and canning shop where they were remelted to form fuel rods (Fig. 3.2). These were machined to size and to prevent oxidation sealed in aluminium cans, a complex task initially involving over 50 operations. The first uranium was cast in mid-January 1948 using metal produced by ICI, and from that October metal manufactured at Springfields was available.[8]

The Windscale Piles

Initially, it was considered that the reactors to produce the plutonium for the bomb's core would follow the design of the wartime United States' Hanford plutonium production reactors. These were fuelled with natural uranium, graphite moderated, and water cooled, each reactor requiring 136,382,700 litres of water per day.[9] Extensive investigations were carried out into possible sites for the reactors, and Arisaig (Inverness-shire) in Scotland was isolated as the only possible location. However, concerns were raised about the safety of the water-cooling system, and alternative cooling methods were considered. These included the use of pressurised gas, although it would require the manufacture of a large, pressurised vessel to contain the core. A third alternative was to use forced air cooling, which British scientists had experience of through the design of the BEPO reactor at Harwell (*see* p. 96). A paper by the AERE Harwell in May 1947 confirmed the viability of using air-cooled piles.[10] An innovation in fuel cartridge design by the addition of fins greatly improved the transfer of heat, allowing the air-cooled option in this large reactor. It was also believed that this less complicated design would be safer, as well as reducing construction time.[11]

With the requirement for vast amounts of water removed, this expanded the options for the location of the piles, although remoteness was still a significant factor. In the late 1940s there were many large government-owned, or recently relinquished, sites that offered large areas of cleared ground and basic amenities that could be allocated to the atomic programme. Initially, the disused TNT factory on the Cumbrian coast at Drigg was proposed as the location for up to three piles.[12] The possible siting of piles with associated ventilation stacks, each originally projected to be 122m tall, so close to the Lake District was met with consternation by the Council for the Preservation of Rural England (CPRE). In a very British response to this proposal, in June 1947 a delegation from the CPRE, including Lady Chorley, met with a representative from the Ministry of Town and Country Planning. A month later the Ministry of Supply announced that the new factory was to be built on the site of another former TNT factory, slightly further to the north, at Sellafield.[13] This is the name of the wider area, and within it some plants were prefixed Windscale, after Windscale Nook on the River Calder, and the later power station was named Calder Hall, after an adjacent farm.

Construction of the first two Windscale piles began in October 1947, and later in February 1949 a third pile was authorised and its foundations started, although it was later cancelled due to American concerns about the supply of uranium.[14] It was a huge civil engineering project and by 1950 employed around 4,750 men, including over 300 architects, surveyors and engineers.[15] This factory had no precedents, and design work often went hand in hand with construction, with drawings rarely released to contractors more than six weeks before the start of work.[16] Its novel requirements also created many manufacturing challenges, such as the thousands of carefully machined graphite blocks needed for the reactor, the production of finned aluminium fuel cans, and the manufacture of the reactor cooling fans, which were the largest yet made in Britain.

Shortly after the end of the war, the Sellafield TNT factory had been sold to the textile firm Courtaulds, which planned to build a rayon factory. Some demolition had started, resulting in a site described as having 'the general appearance of an area which had been bombed or had suffered a major explosion'.[17] A number of the factory buildings, especially in the administrative area, were retained and survived in use for many decades (Fig. 3.3). The project was overseen for the Department of Atomic Energy by the Ministry of Works; construction was split between several organisations. The Ministry of Works' Mobile Labour Force took charge of the main group of buildings, administrative area, parts of the separation group, the hot laboratory, and off-site temporary housing. Costains constructed the piles' foundations, roads and other infrastructure. The principal contractors for the pile superstructures and associated buildings were John Laing and Son Limited, which had constructed the wartime TNT factory. Trollope and Colls were responsible for the separation group and process section, and the third pile that never arose above foundation level.[18] The main steel supplier was Messrs Alex Findley's Works, Motherwell.

By July 1950 construction work on Pile 1 was complete, and it achieved criticality, where enough neutrons were being released to sustain a chain reaction, in October; work on Pile 2 was finished by December 1950 and it went critical the following June (Fig. 3.4).[19] By January 1952 the first irradiated fuel rods were passed to the chemical separation plant, and on 28 March the first billet of plutonium was ready and the first batch was delivered to the Weapons Group in August.[20] Less than two months later, on 3 October 1952, Britain detonated her first experimental atomic weapon on the Monte Bello Islands, 130km off the north-west coast of Australia.

The plant comprised three groups: to the north the Pile Group, in the middle the Separation Group, and to the south the Main Group, which included the administration block, fire station, workshops, stores, a metrological section, and other buildings (Fig. 3.5).

Fig. 3.3
Windscale, Cumbria. The wartime TNT factory headquarters: this building retained the steel frame and roof from a wartime building; new brick cavity walls and windows were added. The senior architect, T. L. Viney, commented that the porte cochère was to 'give a little dignity and emphasis'.
[© Historic England Archive, G5158_3]

Fig. 3.4
Windscale, Cumbria, 20 June 1950. In the foreground work is nearing completion on Pile 2, and to its rear work is underway on the central section of the Chemical Separation Plant. [Historic England Archive (RAF Photography), raf_541_588_f8_0026]

Fig. 3.5
Windscale, Cumbria, 16 June 1952. The completed works: to the bottom right is the Main Group comprising the administration block, fire station, workshops, stores, a metrological section and other buildings. In the middle is the Separation Group with waste storage tanks and the discharge pipeline leading towards the Irish Sea. At the top, marked by the ventilation towers, is the Pile Group with associated blower houses, cooling pond, uranium slug store, police lodge, and workshops. [Historic England Archive (RAF Photography), raf_540_774_psfo_0191]

Fig. 3.6
Windscale, Cumbria. A pile hall under construction showing the steel portal frame.
[Courtesy of the Nuclear Decommissioning Authority, Album 7 P5 1948-1954 0309]

Fig. 3.7
Windscale, Cumbria. The piles shortly after completion; in the foreground is one of the blower houses. Fixed to the left-hand sides of the ventilation shafts are the steel-framed lift shafts that gave access to the air filters, and down which contaminated filters were moved in shielded 'coffins'. Christopher Hinton referred to the piles as 'monuments to our original ignorance'.
[© Historic England Archive (John Laing Photographic Collection), JLP01_10_05336]

The Pile Group

Each pile comprises a large central building that housed the reactor and to either side two blower houses, a fuel store, and dominating the group a tall ventilation shaft capped with air filters. The blower houses are of steel portal frame construction, clad externally with a horizontal glazing band, aluminium louvres and brick infill, and on the roof a glazed clerestory. One of the requirements of the reactor building was the provision of an overhead 25.4-tonne travelling crane. It was decided to form the building from its gantry, providing a cover for construction work, obviating the need to weatherproof the equipment, and giving cover for later reactor operations.[21] This is a huge steel portal-framed structure, 61m long by 27.4m wide and 42.7m tall, with internal cross bracing, set on a massive concrete 3.05m-thick foundation raft measuring 59.74m by 30.48m supporting a weight of 57,914 tonnes (Fig. 3.6).[22] To lessen the demands on the welding shop, most of the steel frame was riveted.[23] The buildings were originally clad in asbestos sheeting with two horizontal glazed bands; these have subsequently been removed and replaced by pressed metal sheets (Fig. 3.7).

The pile, or reactor core, comprised over 50,000 carefully machined graphite blocks forming an octagonal stack 15.2m in diameter and weighing about 2,032 tonnes (Fig. 3.8). This was encased in a concrete biological shield 2.1m thick and 24.3m tall; the reactor roof spanned 15.24m and was 2.6m thick. The core was pierced by 3,440 horizontal fuel channels, which passed straight through the core; on the other two sides were control rods, and on its top, shut-off rods that in an emergency would fall by gravity (Fig. 3.9). Each fuel channel held 21 fuel elements, with a total of around 70,000 in each pile.[24] The fuel elements were about 30cm in length and were tied to a graphite 'boat' (Fig. 3.10). To improve heat exchange, the aluminium 'cans' that encased the uranium fuel rods were given fins. However, it was calculated that there was too much neutron-absorbing material in the pile. The solution was to cut a sliver off each of the fins of the 70,000 fuel elements, an operation led by the deputy works manager Tom Tuohy.[25]

The piles were cooled by a forced-air system with blower houses to each side, each equipped with intake chambers, four large fans and filters (Fig. 3.11). To further prevent heat stress in the

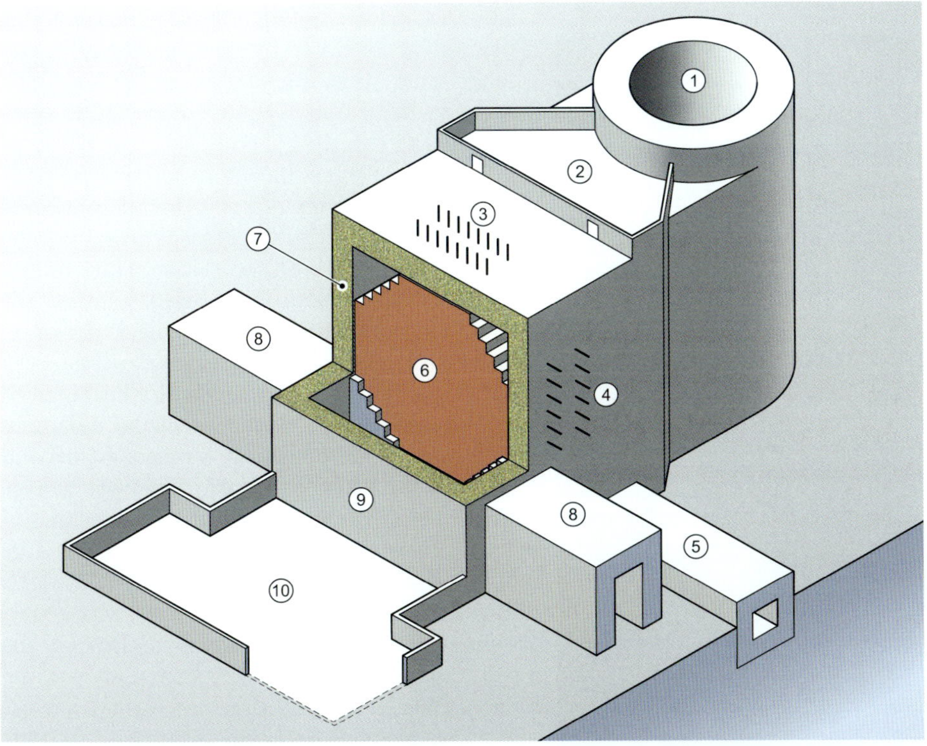

Fig. 3.8
Windscale Piles, Cumbria. Section through one of the piles.

Key
1 Chimney
2 Test house
3 Shut-down rods
4 Control rods – duplicated on the opposite side
5 Water duct
6 Reacting core
7 Biological shield
8 Air duct
9 Charge wall
10 Magazine store

[Redrawn from © Historic England Archive, JLP 01_01_149_27]

Fig. 3.9
Windscale, Cumbria. The pile cap during construction showing the tubes for the shut-off rods.
[Courtesy of the Nuclear Decommissioning Authority, Album 7 P16 1948-1954 1327]

Fig. 3.10
Windscale Piles, Cumbria. Fuel element with a finned can tied to a graphite boat. Ten different types were designed, six of which were used between 1951 and 1957. During their operation the two piles used around 750,000 fuel elements.
[Courtesy of the Nuclear Decommissioning Authority, P-56-762-R-AEA]

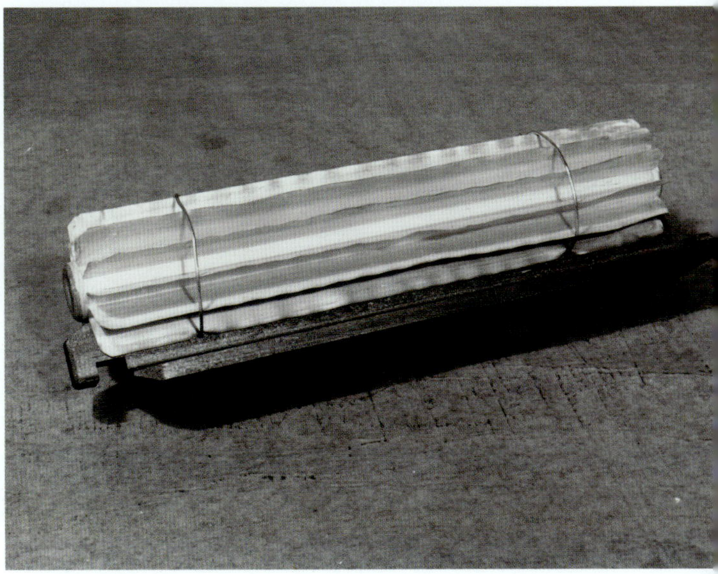

surrounding concrete biological shield wall, the reactor was enclosed by steel shield plates, and to the rear of these, insulation containers which acted as an air duct. To prevent any dust entering the system, the concrete wall was enclosed in aluminium sheeting. On the rear of the reactors were the discharge plates, and connected to these was a series of guides, known as 'curtain rails', that were used to direct aluminium 'spears' that were in turn used to raise the stainless steel sniffers used for leak detection (*see* Fig. 3.13).

In operation, the fuel elements were received from Springfields and placed in the magazine store before being put in the magazine lift and taken to the loading platform. This was a large steel lift platform, reputedly derived from an aircraft carrier hoist, that sat within the charge hoist well; its level was indicated by a low-tech finger pointer. From this lift, access was gained to the charge holes that passed through the charge wall and into the reactor. Between the charge face wall and the reactor was an air duct that was bridged by a tube before loading took place. The fuel channels were arranged in groups of four and at the centre of each group there was a separate channel for isotope rods (Fig. 3.12). A charging machine, later discarded, was used to load the reactor with fuel elements, as well as materials to produce isotopes. Further charging holes on top of the reactor were also used for loading various materials to study the effects of radiation.

To defuel a channel, the boats, with the fuel elements tied to them, were simply pushed out of the opposite end of the channel, where an oval opening allowed the elements to break away from one another (Fig. 3.13). The fuel element, still tied to its boat, then fell by gravity into a rail-mounted trolley that ran along a water-filled channel to a large prestressed concrete water-filled cartridge cooling pond (Fig. 3.14). This was 106.6m long, 24.3m wide, and 6.7m deep, 5.5m of which projected above the ground. It was divided into two by a central wall, and each of the tanks was asphalt lined, with a travelling crane above.[26] In the processing bay of the pond a submerged press was used to strip the aluminium cans from the fuel elements which were then raised to the top of the Chemical Separation Group (*see* p. 36).

Fig. 3.11
Windscale, Cumbria. An interior view of one of the blower houses showing the blowers that blew air through the reactor that lay to the right. To the left is the raised control room. [Courtesy of the Nuclear Decommissioning Authority, Album 7 P13 1948-1954 3334]

Fig. 3.12
Windscale, Cumbria. View through one of the holes in the charge face showing the arrangement of the fuel-element channels and the central channel for an isotope rod. [Courtesy of the Nuclear Decommissioning Authority, Album 7 P33 1948-1954 3865]

The ventilation shafts

One feature of the construction of the piles was flexibility in planning, to accommodate changes resulting from increased understanding of the design of reactor components. One example of the benefits of this approach was in the construction of the ventilation shafts. They were originally designed to be about 122m tall, but with the provision to rise to 137m if wind tunnel tests showed that the extra height was needed to disperse the vented air. Late in 1948, after John Cockcroft had visited the United States, it was determined that air filters needed to be fitted. The shafts had already risen to 21m, and the only option was to position them on top of the vents. This involved the construction of steel, concrete and brickwork weighing 2,540 tonnes at a height of over 122m.

When completed, the ventilation shafts stood to 126.4m with an internal diameter of 13.4m tapering to 10.6m at the top. At their bases, where they also formed part of the biological shield, they were 1.5m thick, tapering to 0.30m on the underside of the galley. Internally, the shafts were lined with aluminium alloy plates with thermal insulation to the rear. To remove

Fig. 3.13
Windscale, Cumbria. The discharge face showing the oval holes that allowed the fuel elements to separate and the 'curtain rails' that acted as guides for the sniffers used to detect burst cartridges.
[Courtesy of the Nuclear Decommissioning Authority, Album7 P31 1948-1954 3350]

Fig. 3.14
Windscale, Cumbria. The open cooling pond between the Windscale Piles; above the pond is a travelling crane for moving the fuel elements and to the right are empty skips.
[Source: Historic England Archive, JLP01_09_740385]

the used filters, which might contain radioactive material, they were first transferred to a water trough in the gallery before being moved in lead-lined containers, or 'coffins', to the lift shaft, a steel-framed and asbestos sheeting-clad structure fixed to the shaft's exterior.[27]

The Windscale fire, October 1957

The Windscale fire of October 1957 was the worst nuclear accident Britain has experienced, resulting in the release of radioactive substances, including iodine-13 which was detected in local milk.

When the Windscale Piles were designed, relatively little was known about the behaviour of graphite under long-term irradiation. One phenomenon that was recognised was the 'Wigner effect', named after the Hungarian émigré physicist working in the United States, Eugene Wigner, who had predicted that under the effects of heat and irradiation that graphite would display some unusual properties. Initially, it can be expected to expand because of heating, contract due to the combined effects of heat and irradiation, then expand with Wigner energy. This stored energy is created by neutron bombardment knocking carbon atoms from their normal position in the crystal lattice of graphite. It was acknowledged that if left unchecked this phenomenon might result in a reactor fire.[28]

Unforeseen temperature rises had been noted in the Windscale Piles as early as 1952, and subsequent investigations concluded that they could be attributed to the release of heat caused by Wigner energy. A trial in January 1953 determined that this energy could be removed by allowing the temperature of the graphite to rise through nuclear heating, a process known as annealing. Based on monitoring conditions within the reactors, the main cooling air blowers were periodically turned off and the temperature in the pile allowed to rise by nuclear heating. After the initial experimental release, a further 15 annealing exercises took place between then and July 1957.[29]

The ninth anneal planned for Pile 1 in early October 1957 appeared to be a routine operation, a procedure that had been carried out many times before. The only unusual features were a slightly longer interval between annealing and greater variations in the pile's internal characteristics. On Monday 7 October the blowers were switched off, and as the nuclear heating began, the temperature started to rise, with the control rods being manipulated to direct the heating within the pile (Fig. 3.15). The anneal appeared to have worked as expected, and in the early hours of the next day the pile was shut down. A few hours later the pile temperature appeared to be static or falling suggesting that the Wigner energy release was no longer self-sustaining, and due to varying conditions within the pile might leave areas unannealed. It was therefore decided to restart the reactor to attempt a second anneal, a procedure which had been done twice before. This was carried out in the early evening. For the next few hours nothing indicated that anything was amiss, until Wednesday afternoon when rising temperatures were detected. Standard operating procedures were followed, and the temperature began to fall. The following day further temperatures rises were detected, and attempts were made to reduce these with the shutdown fans and main blowers. Traces of radioactivity were also detected in the stack and other sensors on site. On Thursday afternoon the pile staff began to report their concerns to their managers. Suspecting that the cause might be a burst cartridge they started to prepare to discharge the affected channel. When the charge plug was removed a red glow could be seen, which was repeated after further plugs were detached (Fig. 3.16). To create a firebreak in the pile, a team of eight workers commenced discharging fuel elements from around the fire into the central cooling pond. They then moved on to pushing out fuel elements from the channels which were on fire. By the early evening the deputy works manager, Tom Tuohy, had been called from home and, standing on top of the reactor roof, confirmed the extent of the fire. Choices for cooling the pile were limited and risky: turning the blowers on might fan the flames, and to introduce water might result in an explosion. Early on Friday morning, however, it was decided there was no choice but to go with the latter option and the establishment's dedicated fire brigade began to slowly pump water into the pile. The water pressure in the hoses was gradually increased and the remaining fan shut down. By Saturday the pile was cold.[30]

The immediate danger of a larger reactor fire, or even a water-induced explosion, had been averted by the bravery of the site staff. Beyond the factory fence, iodine-131 was detected, leading to the widespread destruction of milk supplies in the vicinity. A Board of Inquiry under William Penney was convened, supported by technical evaluations at other United Kingdom Atomic Energy Authority (UKAEA) sites.[31] The

Fig. 3.15
Windscale, Cumbria. Reactor control room. Instruments in the room monitored neutron flux, blower speeds, control rod positions, and temperature through thermocouples in the core. The room was also equipped with alarm signals and trips. In an emergency a master switch could bring all the safety rods down in one movement. Another switch could close the main blowers.
[Source: Historic England Archive, JLP01_01_149_19]

Board reported very quickly, on 26 October. It attributed the main cause of the fire to the second annealing and a failure of one or more fuel cans leading to an oxidisation of the uranium fuel rods. Another possibility was that the fire started in a lithium-magnesium cartridge used to produce tritium, a vital substance for the hydrogen bomb.[32] Lorna Arnold, the UKAEA official historian, concluded the cause, or causes, are still disputed.[33] The operational staff had been put in an impossible position: to run the pile without releasing the Wigner energy would have probably led to a fire, and a similar risk was presented by annealing.[34]

The causes of the fire were also deep-rooted and organisational. Since the end of the war the industry had grown very rapidly and there was a constant struggle to recruit enough qualified scientists and engineers. The programme had always worked to very tight deadlines, first to deliver devices for overseas weapons trials and subsequently to produce material to build up an operational stockpile. It had then been expanded to encompass the development of hydrogen bombs, which required other novel materials. The world's first civil nuclear power programme had been initiated in parallel with weapons development. This in turn led to ongoing research into a variety of reactor types. With hindsight, perhaps more operational research should have been undertaken into the condition and operating characteristics of the piles. They, however, represented an obsolescent system, which would not be replicated in any future scheme.[35]

Pile 2 ran for another few days until the 17th when it too was shut down. The cost of the

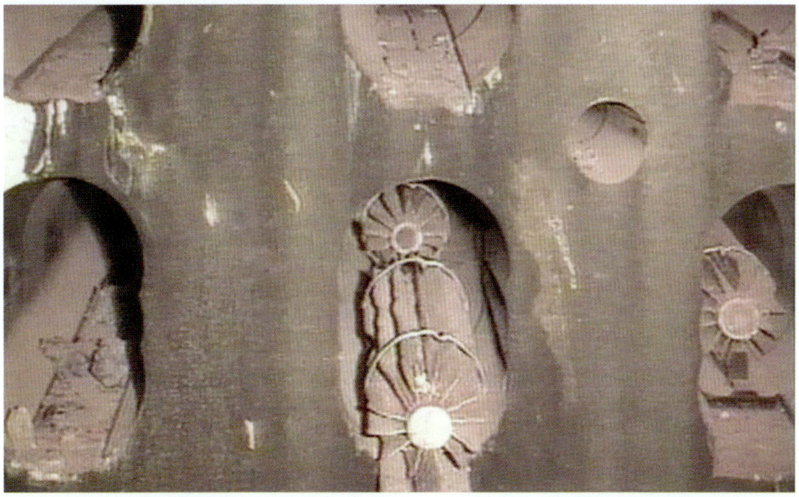

Fig. 3.16
Windscale, Cumbria. Section of Pile 1 showing part of the area affected by the fire, fuel elements with finned aluminium cans tied to graphite boats trapped in fuel channels.
[Courtesy of the Nuclear Decommissioning Authority, Historic Piles CFR]

necessary safety modifications to the piles made resuming operations uneconomical, and in autumn 1958 it was announced that neither would restart. Even if they were restarted, they would have a relatively short operational life, as within a few years a new generation of reactors would be in service. Work began to defuel the piles, except around the fire, which was left.[36] By 1960, the buildings and ground around the piles were decontaminated and Pile 1's blower house was reused to house the HERO reactor and other facilities to support the Windscale Advanced Gas Cooled Reactor (*see* Fig. 6.7).

Beyond the immediate technical and organisational issues, the accident presented political challenges for the industry. Only a few days before the accident the Soviet Union had successfully demonstrated her technical prowess with the launch of the world's first artificial satellite – Sputnik. How would the accident affect relationships with the United States and the desire to re-establish nuclear cooperation, especially in the defence field? Would there be any consequences for public support of the civil nuclear power programme? Internally, the site's operator, the UKAEA, was restructured, with a more prominent role given to its health and safety branch. As the government's virtual monopoly on the design of nuclear facilities through UKAEA lessened, more formal regulation was exercised through the 1959 Nuclear Installations Act, which in 1960 established the Nuclear Safety Division within the Ministry of Power as the Inspectorate of Nuclear Installations. A decade later this was followed by the National Radiological Protection Board. UKAEA and the later independent Atomic Weapons Establishment retained their Crown immunity and operated outside this regulatory regime until the 1990s.

The Chemical Separation Group

Located to the south-west of the piles, the main function of the Separation, or Chemical, Group was to extract uranium and plutonium from the irradiated fuel cartridges and then separate these elements. The development of the novel solvent extraction process was overseen by Dr R. Spence. Due to the absence of suitably shielded facilities in the United Kingdom, work was initially undertaken at Chalk River, Canada,

Fig. 3.17
Windscale, Cumbria. The Chemical Group with B204 to the rear.
[Source: Historic England Archive, JLP01_12_015]

Fig. 3.18
Windscale, Cumbria. Uranium purification plant, showing its portal frame construction. Picture probably taken in the early 1950s.
[Source: Historic England Archive, JLP01_12_016]

assisted by colleagues from Harwell and Risley, aided by staff from ICI. Some experience of extracting plutonium had been gained with microscale work at Chalk River using irradiated uranium fuel rods, or slugs, from the Clinton Pile. At Chalk River a pilot extraction tower was built by a Harwell team led by C. R. Nicholls, using irradiated slugs from the plant's NRX (National Research Experimental) reactor.[37] In common with the Piles Group, the engineers in charge of the design of the Chemical Group lacked precise information on the plant and processes to be housed within. The 28,449-tonne main process building, B204, was constructed on a substantial concrete mat 44.2m by 22.9m, laid to a honeycomb pattern to reduce its dead weight (Fig. 3.17).[38] On this a 60.96m-tall reinforced concrete shell with projecting heavily shielded chambers was built in which the process plant would be installed (*see* Fig. 3.4). Holes were left in the wall into which pipework and control equipment could be inserted and later encased in concrete. Cast beams spanning the internal space would subsequently support tanks and pipes. The large concrete shell was surrounded by chequer-plate operating floors at 4.6m intervals, on which control and non-radioactive equipment could be placed. The whole was enclosed by a large portal steel-frame structure, clad in horizontal bands of asbestos sheet and patent glazing (Fig. 3.18). Towering above the building was a vent stack, its top 124m above ground level. Kenneth Jay, a contemporary writer on the nuclear industry, commented it was a 'plant built on a foundation of inspired guesswork'.[39]

Due to the radioactivity of the processes, the design of the chemical plant posed particular challenges, and once it was operational no one would be able to enter the concrete shell to carry out remedial or maintenance work. To remove upkeep requirements, it was built without pumps and valves. To resist the acids, solvents and radioactivity, a new type of stainless steel

Fig. 3.19
Windscale, Cumbria. Working deck of the primary separation plant. Picture probably taken in the early 1950s.
[Source: Historic England Archive, JLP01_12_014]

was required, and a new plant was built for its manufacture. Within the concrete shell the separation plant comprised a series of pipes, columns and tanks (Fig. 3.19). They were assembled with extreme care; each part was numbered, and the 50,000 welds were visually inspected with either X-rays or gamma rays. Such was the care taken that each joint had to be certified as correct before another piece was attached to the next component.

In operation, slugs taken from the cooling pond, minus their aluminium cans, were dissolved in nitric acid and the resulting liquor poured into columns to separate out the plutonium. Around one pound (0.45kg) was recovered from every ton of discharged containers.[40] The main products of the plant were an impure solution of plutonium and uranium depleted in the 235 isotope, dirty solvent, and a highly radioactive solution of fission products. The impure uranium was purified and returned to Springfields where it was converted to uranium hexafluoride before being shipped to Capenhurst (Cheshire) for enrichment (*see below*). The plutonium solution was moved to a smaller building for refining. Again, construction of this building began without detailed knowledge of the process plant to be installed. However, the quantities of material to be processed were smaller and plutonium is comparatively less radioactive, and thus required less shielding. To reduce fire hazards the plant operated in an atmosphere of inert gas. To speed up construction, sections of the plant were constructed from quartz glass panels encased in Perspex. This provided more flexibility in construction, and during use good visibility of the process, although it was noted that if the process had been better understood, stainless steel might have been preferred.[41]

Effluent from the chemical plants and other radioactive waste from the factory were classified as high, medium and low. High-level waste was stored indefinitely on site in tanks, acknowledged at the time as a problem for the future. Medium-level waste was stored while its radioactivity decayed and after a few years was mixed with low-level waste for discharge via a 3.22km pipeline into the Irish Sea (*see* Fig. 3.5). Studies were undertaken to minimise any negative effects, but in later decades the expedients taken related to the treatment of waste associated with the delivery of plutonium for the first atomic bomb trials would bring notoriety to the Sellafield site.[42]

Uranium enrichment, Capenhurst, Cheshire

The third new factory built by the Production Division was at Capenhurst (Fig. 3.20). Initially its principal function was to enrich the uranium depleted by its time in the Windscale reactors, which was returned to Springfields where it was converted to uranium hexafluoride, or hex, and then conveyed to Capenhurst for processing. Later, enriched uranium was required for use in weapons development and new types of reactor fuels, including submarine reactors. Capenhurst was a wartime Royal Ordnance small arms ammunition factory; as such, it was a relatively recent factory with basic infrastructure and a direct connection to the Chester-to-Birkenhead railway line. This location also provided access to water from the River Dee for cooling, electricity from local power stations, and an urban catchment area from which its workforce might ultimately be drawn. During its construction phase, however, hostel accommodation for up to 850 men was provided 7km away at Sealand, Deeside, and personnel from the Ministry of Works' Mobile Labour Force were deployed.[43] The area was relatively flat, with surrounding farmland offering the opportunity for further expansion. Once again environmental concerns were raised as the area was identified as possible greenbelt for the nearby town of Birkenhead. To allay concerns, Christopher Hinton met with the local authority to reassure them about the appearance of the buildings.[44]

Research into the enrichment of uranium to increase the proportion of the more fissionable lighter uranium-235 isotope had begun during the Second World War, supported by the MAUD Committee tasked with exploring the feasibility of an atomic bomb. As described in the previous chapter, British investigations into diffusion technology were curtailed by the August 1943 Quebec Agreement. After the war, resumption of this research was less urgent than the production of plutonium. Interest restarted in late 1947 with basic research at Harwell and an ICI pilot plant at Springfields prior to the establishment of the development laboratory at Capenhurst.

Work on the new factory began in March 1950 and production commenced in 1953. Buildings were laid out to the west of the wartime factory following its orientation parallel to the railway line. Perhaps to fulfil Hinton's

promise to the local authority, the main administration building, double storeyed under a mansard roof with dormers, was set askew to the main alignment, effectively screening some of the factory from Capenhurst village (*see* Fig. 3.20). Established trees were also left in position. In common with the other building projects, work began before the final plant specifications were complete. The main process building was a large, utilitarian, rectangular, six-bay structure supported on a steel lattice-girder framework and clad in corrugated sheeting. Internally, this provided a large open-floor area for the repetitive steel boxes that housed the compressors and pipework. Above each unit was a working floor with control and monitoring equipment; advanced automation methods allowed the plant to be supervised by a comparatively small workforce. Diffusion was a single physical process and the basic unit within the plant comprised a pump or compressor, a membrane and connecting pipework and a cooler system to remove heat produced by compressing the gas. To disperse the heat the plant was initially equipped with two cooling towers. At each stage only a minimal amount of enrichment took place, and to achieve the desired concentration the process was run as a continuous operation with consecutive units performing the same task. The construction of the plant posed many challenges, including sealing all the plant to operate within a vacuum, the handling of highly corrosive uranium hexafluoride, and the necessity for absolute cleanliness. For this purpose, a workshop in the process area was dedicated to receiving and cleaning plant

Fig. 3.20
Capenhurst, Cheshire, August 1952. The main contractor for the work was Trollope & Colls Ltd. At this time, it was one of only two uranium enrichment plants in the western world. [Historic England Archive (RAF Photography), raf_540_836_sffo_0032]

Fig. 3.21
Capenhurst, Cheshire, September 1968. The greatly expanded gas diffusion plant. The 11 cooling towers are indicative of the factory's colossal power consumption and waste heat generated by this process.
[Courtesy of the Nuclear Decommissioning Authority, OP36272 18991]

components from the manufacturers before installation. Towards the end of the process, the hexafluoride was separated from the uranium by freezing, using liquid nitrogen produced within the process area.[45]

After a relatively short period, enough enriched uranium had been produced for the weapons programme. In late 1962 it was announced that manufacture would be drastically scaled back to provide for limited civil requirements and to maintain expertise, a decision implemented by late 1963. However, a few years later, the adoption of advanced gas-cooled reactors (AGRs) (*see* p. 202) created a new requirement for enriched uranium for use in reactor fuels. In late 1965 it was announced that £13.5 million was to be invested in modernising the works (Fig. 3.21). The familiar gas diffusion technology was retained as it was felt that centrifugal separation technology was as yet inadequately developed.[46] However, by September 1977 centrifuge technology developed at Capenhurst was sufficiently advanced to allow the opening of the first centrifuge hall. This process used about one-tenth the energy input of the diffusion process, and the cooling requirements could be met by small roof-mounted plant. The new technology led to the eventual closure of the diffusion plant and demolition of the 1950s cooling towers.[47] Now managed by Urenco, the factory remains in operation and has added important new facilities to process and safely store the by-products of uranium enrichment.

Box 3.1 New communities

In contrast to nuclear facilities in many parts of the world, the men and women of the new 'atomic' communities in the United Kingdom were generally well integrated into their local areas. Geographically, the United Kingdom was too small to create the closed nuclear cities of the Soviet Union, or to set aside the huge ranges of the United States' interior. Nor was this the desire of the leaders of the new industry, who wished to attract and retain highly skilled staff and encourage them to settle close to their places of work. In the early post-war years, housing was scarce and the prospect of accommodation could be an attractive incentive for personnel to work at a particular establishment. Where housing was provided by government departments, it was analogous to munitions workers' and armed services' estates, where the type of accommodation offered mirrored that of the establishment's hierarchy, usually that of the husband's grade. In contrast, where local authorities provided housing, it tended to be egalitarian in form. In appearance, permanent housing was conservative, generally in brick with pitched tiled roofs, usually set in open estates. Reflecting post-war building material shortages, prefabricated aluminium bungalows, concrete Airey houses, steel houses by the British Iron and Steel Federation (BSIF), and timber houses are also encountered. These traditional-looking estates differed from, for example, the Atoomwijk, close to Mol, Belgium, where the lead architects Jacques Wybaum and Jacques Thiran created a Modernist woodland settlement for the scientists, engineers and foreign visitors in a vision of a new atomic world.

Initially, the construction of the research establishments and later nuclear power stations required large itinerant workforces; typically, a few thousand workers might be employed on a project for a few years. Comparable to Victorian navvy camps, many acquired a reputation for hard living and rowdiness. Recently abandoned munitions workers' and military camps were ideal for accommodating large numbers of single men. At Dounreay, airfield hutting and an army camp at Thurso were used to house the construction workers (Fig. B.3.1.1). Later, at Winfrith, former RAF Warmwell (Dorset) was acquired, and at Trawsfynydd power station (Gwynedd) the construction team used a recently vacated army camp. Elsewhere, in addition to temporary offices and workshops, accommodation camps for construction workers were built. Around various facilities these may remain visible as areas of flat open ground.

Many of the early establishments were built on recently vacated airfields or munitions works, and one of the attractions of these sites was their existing infrastructure, including accommodation and social facilities. Harwell had 80

Fig. B3.1.1 Dounreay, Caithness. Martinet Block, May 1954. Abandoned airfield Nissen huts soon after their acquisition by UKAEA. [© Historic England Archive, G5164_1]

good-quality RAF married quarters, graded according to rank; John Cockcroft and his family were allocated the former commanding officer's house. However, for a rapidly expanding workforce that peaked at about 6,000 in 1959, this housing provision was far from adequate.[1] The former RAF officers' mess continued to offer quarters, and some of the barracks were converted into hostels for single and junior staff, later supplemented by the Rush Common hostel in Abingdon (Oxfordshire). To the north and south of the Harwell airfield in the summer of 1946, 200 Mk IV prefabricated aluminium bungalows were erected by German prisoners of war; many remained in occupation until the late 1980s (Fig. B3.1.2, *see* Figs 4.1 and 4.19).[2]

Offsite, several country houses were leased, including Cosenor's House and Fellows Close near Abingdon and the imposing Palladian Buckland House, listed Grade II. These were used to accommodate more senior staff and also served as conference venues. Less grand were lodgings at the former Royal Ordnance Factory camp at Grazeley Green, adjacent to Burghfield (Berkshire), where travel between the two sites entailed a daily 60km round trip by bus. In the surrounding Oxfordshire towns, the Ministry of Works built new estates in Abingdon, Wantage and Wallingford, totalling 1,100 houses; a further 500 houses were allocated to AERE by local councils. To the north of Abingdon, the Ministry of Works acquired the Fitzharris Manor House, a 16th-century house and grounds. Development here was reserved for more senior staff and reflected in the generous spacing of seven houses per 0.4ha. This gave room for open green spaces and wide grassy verges; mature trees were retained from the estate's gardens and more planted. In total, 140 semi-detached houses were built, half with three bedrooms and, with a special dispensation from the Ministry of Health exercising its post-war responsibility for housing, the remainder were given four bedrooms (Fig. B3.1.3).[3] Despite continuing petrol rationing, most were also provided with single garages. On 2 December 1952 it was reported to the House of Commons that the Ancient Monuments Board had been consulted about the manor house, and due to its poor condition consented to its demolition.[4] The houses in Wantage, including some BSIF steel houses, were generally for lower-graded staff and built without garages. At Aldermaston, only temporary wartime airfield hutting was available, and a large housing estate was built immediately south of the site.

Risley (Warrington Unitary Authority) was close to urban conurbations of the North West and was able to draw some staff from their home areas. But housing was required for people from outside the area, including specialists and the critical senior staff to head departments. As at

Fig. B3.1.2
Harwell, Oxfordshire, August 1947. One of the estates of aluminium prefabricated housing built adjacent to the airfield, known as Prefabria.
[Source: Historic England Archive, P_H00214_001]

Fig. B3.1.3
Stanford Drive, Abingdon-on-Thames, Oxfordshire. AERE housing soon after completion in May 1948. This view shows the generously spaced housing estate with a retained Wellingtonia from the old estate gardens. Here this 'atomic' community was lit at night by newly installed gas lights.
[Source: Historic England Archive, P_H00262_016]

Abingdon, it was decided to allocate senior staff to a discrete estate, again with larger houses with garages. This was built to the south of Warrington at Appleton Hall. Here security appears to have been one reason for this distinction. Many of the new houses shared telephone connections, party lines, potentially allowing conversations to be overheard. It was also felt different grades would be more comfortable socialising separately.[5] Accommodation was also available a few kilometres to the north at the former HMS *Ariel* camps at Newchurch Hall and Little London. Two hundred three- and four-bedroom semi-detached houses were also built at Culceth.[6]

Windscale (Cumbria) was also able to use remaining wartime factory accommodation for construction workers and single staff (Fig. B3.1.4). A temporary construction camp was also established at Nethertown and hostels at Greengarth and Mirehouses.[7] Here, too, new estates were required, and to the south at Seascale, in September 1947, an initial request was made for 200 houses, split between 10 Type A detached, 75 Type B four-bedroom semi-detached and 115 Type C three-bedroom semi-detached houses (Fig. B3.1.5). A year later the number of requested houses rose to 304, and by 1958, 918 houses had been built.[8] At Egremont, terraces of aluminium houses were constructed.[9] Whitehaven Corporation was asked to provide 360 houses, resulting in a new estate, Mirehouses, built between the town and St Bees with Ministry of Supply assistance.[10] It was not the most attractive choice, however, given initial tensions with the local population, its distance from Windscale, and fumes from a local chemical plant.[11] Others might take local private lodgings or buy on the open market. In 1955 the deputy works general manager, Tom Tuohy, bought an architect-designed, early 20th-century villa at Beckermet, now listed Grade II.[12] Differing priorities in the Ministry of Supply and Ministry of Works in the allocation of labour and materials was a constant frustration for Christopher Hinton, Deputy Controller of Production, as delays in the provision of residential accommodation and social facilities impeded his ability to recruit staff to maintain challenging production schedules.[13] At Seascale, it was felt that delays in the construction of the primary school was having an adverse effect on the recruitment of personnel. To remedy this situation, workers were transferred from factory

Fig. B3.1.4
St Bridget Beckermet, Copeland, Cumbria. A characteristic bed-sitting room in the hostel for Windscale staff. To create a home from home, personal items have been added to the utility-standard Ministry of Supply-provided furnishings. [Source: Historic England Archive, JLP01_01_149_20]

Fig. B3.1.5
Eskdale Avenue, Seascale, Cumbria. 1950s housing built for the workers at Windscale, illustrating the estate's wide curving roads and mixed housing types. The use of stone panel detailing in the porch was a small concession to the local vernacular. [© Historic England Archive, DP081121]

Fig. B3.1.6
Seascale Primary School, Croft Mead Road, Cumbria. A modern school of flexible design, its completion was crucial for attracting workers to this relatively remote part of the country. [© Historic England Archive (John Laing Photographic Collection), JLP01_01_108_64]

construction to the school to ensure its completion by early 1951 (Fig. B3.1.6).

Winfrith (Dorset) was in a rural area with relatively little local accommodation, and from the outset the policy was to disperse the workforce among the local towns and villages, including Dorchester, Wareham, Poole and Bournemouth. In the latter two, UKAEA bought 153 houses, and a further 127 were built in nine other areas. To the south of Dorchester, the town council constructed 120 houses on what was to become the Manor Park Estate. UKAEA also acquired the Durley Hall Hotel in Bournemouth for use as a hostel and as visitor accommodation.[14]

The most isolated atomic research centre was at Dounreay (Caithness) in the far north of Scotland. Here the main accommodation was 14km to the east in the market town of Thurso, where four estates were built at Pennyland, Mount Vernon, Castlegreen and Ormilie. Smaller estates for senior staff were also constructed at Scrabster and Castletown. These generally comprised detached, or large semi-detached, houses with distinctive chimney stacks in local stone. In total, between 1955 and 1963, 1,007 houses were constructed for Dounreay 'Atomics', and the population of Thurso tripled from around 3,000 to over 9,000.[15] The residences here displayed a great assortment of designs, including detached housing for senior staff, three- and four-bedroom semi-detached housing, blocks of flats and hostels, and used a range of building materials (Fig. B3.1.7). The estates followed best contemporary practice with a relative low house concentration of nine or ten houses per 0.4ha, thought by some commentators to be too low a density. They were open and uncluttered, with underground services, an absence of boundary walls, and the extensive use of rear garage blocks separated pedestrians and vehicles (Fig. B3.1.8). To reduce the cost of groundworks required because of the sloping topography of the area, semi-detached housing was favoured, which could be placed on smaller platforms. The estate designs were drawn up by a local architecture practice, Sinclair MacDonald & Son of Thurso, led by Hugh Sinclair, according to specifications set by UKAEA. Construction was put out to tender, the successful firms then being responsible for the houses' outward appearance, subject to comments from the local authority's architecture office at Risley. A novel feature of the Thurso estates was the extensive use of timber, which after the end of rationing had only recently become more widely available. In total 156 timber-built houses were constructed to a variety of designs produced by W. J. Simms & Co. and Cooke Ltd, Nottingham (Fig. B3.1.9), including single- and double-storey examples.[16] Close to Thurso railway station was Ormlie Lodge Hostel, in origin a large Victorian villa. Here a conventional three-storey extension was built to the rear in brick, while to the front of the house R. A. Brocklesby, architect to UKAEA, designed a three-storey extension. This included elevations in local stone and external red cedar boarding brightened with red, blue and green window boxes. This hostel provided 72 individual rooms for single staff and on the ground floor a large open dining room.[17]

The research establishments were large enterprises employing a few thousand people in each. To meet safety considerations, most of the new atomic power stations were placed in remote rural areas and initially might employ 300–450 people. While many station workers might be recruited locally, further housing was required for incomers. Bradwell power station in Essex, with an operational staff of around 450, required new estates at Latchington, Southminster and Tillingham.[18] Similarly, the medieval village of Stogursey (Somerset) was greatly expanded to accommodate some of the workforce from Hinkley Point. In appearance there was little to distinguish these developments from contemporary local authority housing estates.

A benefit for the newcomers and existing residents was often an improvement in educational opportunities. To meet the needs of the increased population at Thurso, a new secondary school, designed by Basil Spence, was completed by 1958. To ensure a well-trained technical work force, apprenticeships were set up within the research establishments, often supported by local technical colleges, as at Whitehaven (Cumbria) and Thurso. At Winfrith, an apprentice school was founded, later also serving apprentices from Risley. To accommodate trainees from outside the area a hostel, Egdon Hall, for 62 boys, was built about 25km away in Weymouth. The buildings were mainly single-storeyed, providing two, four and six berths and communal facilities; a single double-storey building was occupied by the warden and stores.[19] Today, some of the buildings remain incorporated into a garden centre.

Fig. B3.1.7
Queens Terrace, Thurso, Caithness. Springpark estate built by the local housing authority, under construction in May 1954. The end house facing the River Thurso and the old town is faced in local Caithness stone.
[© Historic England Archive, G5169_6]

Fig. B3.1.8
Thurso, Caithness. A typical carefully manicured and carless UKAEA estate.
[© Wayne Cocroft]

Fig. B3.1.9
Sweyn Road, Thurso, Caithness. Pennyland estate, examples of timber housing and to the left a three-storey block of flats.
[© Wayne Cocroft]

In addition to housing, social centres and sports grounds were created for the employees. In many areas the new atomic workers represented a large influx of a highly educated population; in Abingdon, AERE staff made up a quarter of the town's population of 13,000. Rather than standing aloof from the townsfolk, the new residents were active in local clubs and societies, voluntary groups and churches, often taking leading roles through committee membership, a pattern found elsewhere.[20]

Today, most of the temporary wartime and early post-war accommodation has been cleared. However, the generally well-built, generously spaced estates of brick houses discussed above, now in private ownership, remain as desirable homes and a largely unrecognised legacy of the atomic age.

Armament Research Department, Fort Halstead

While teams laboured in the north of England to produce the necessary fissile material for a bomb, other groups in the south worked to perfect the technology necessary to produce the precisely timed explosive-induced implosion necessary to create a nuclear chain reaction. The work was led by the Armament Research Department (ARD), which had originated in the Research and Design Department, originally based at the Royal Arsenal, Woolwich. During the late 1930s it had acquired Fort Halstead (Kent), a late 19th-century mobilisation centre (in essence a fortified depot or redoubt) to the north of Sevenoaks, which was redeveloped with buildings of a quality commensurate with a permanent establishment.[48]

On his return from the United States on 1 January 1946, William Penney, physicist and veteran of the Manhattan Project, was appointed as Chief Superintendent of Armaments Research (CSAR).[49] In January 1947, after the formal Cabinet decision to develop an atomic bomb, he was an obvious choice to head the British bomb project. The ARD would lead the project, but to maintain secrecy a separate section called Basic High Explosives Research (BHER), usually abbreviated to HER, would be established. This appears to have been constituted in early 1947, and by April Penney was beginning to consider the facilities and staff he would require at Woolwich and Fort Halstead.[50]

At first, the accommodation needs of the project were modest, and in keeping with contemporary austerity measures current ARD sites were chosen, buildings modified and comparatively few new structures built. Fort Halstead offered a secluded location, with relatively new buildings and a section for handling small amounts of explosives. The link between this project and the fort was a closely guarded secret, to the extent that even within government all orders were routed through the research establishment at Harwell until 1949.[51] That said, an article entitled 'Dr Penney plans new weapons in 1807 invasion fort' in the *Daily Express* in 1947 by the investigative journalist Chapman Pincher might have alerted even less than astute intelligence agents to the connection. Although correct in linking Penney with the work, Pincher was mistaken about the date of the fort.[52]

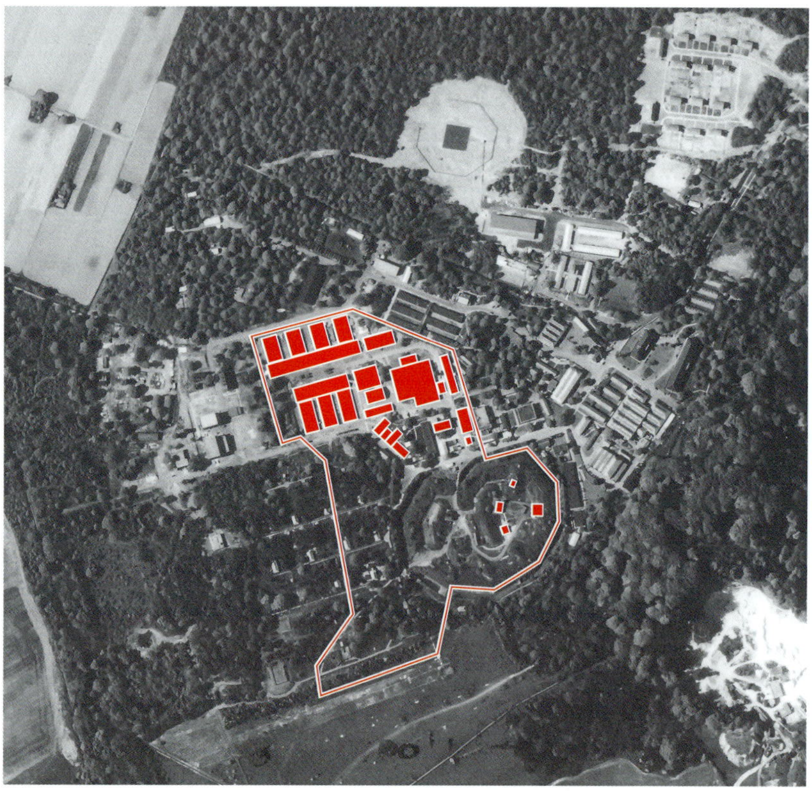

Fig. 3.22
Fort Halstead, Kent, 15 May 1952. The ramparts of the Victorian Mobilisation Centre are to the bottom right. The red line denotes the enclave occupied by the High Explosives Research team. Buildings associated with early atomic work added to this area between 1947 and 1952 are shown in red.
[Historic England Archive (RAF Photography), raf_540_731_rs_4075]

To further guard the confidentiality of the work, Penney's team was housed within a fenced enclave enclosing the Mobilisation Centre and a group of buildings to its immediate north (Fig. 3.22). William Penney's main office remained in the late 1930s CSAR headquarters offices building, which was located just inside the main gate. Within the HER enclave were the 19th-century fort and several wartime buildings, to which further purpose-built structures were added. One of the responsibilities of the electronics team at the fort was the development of the detonators that were required to simultaneously detonate each of the 32 explosive lenses surrounding the core to initiate the nuclear chain reaction. These were perfected and tested at the fort, while their manufacture was carried out at a separate establishment adjacent to Woolwich Common (*see* p. 54).

To assist with this work, the HER area incorporated several wartime explosives filling sheds to its west, and it was probably also at this time that a breach was made in the western defences of the Mobilisation Centre. This allowed for the safe transfer of trial assemblies to the test chambers constructed within (Fig. 3.23). These were sited on its eastern side, and use was also made of its subterranean casemates for experimental work. The first detonation laboratory was designed in July 1947, just a few months after

Penney had been given the task of developing the bomb (Fig. 3.24).[53] Internally, this has a simple E-shaped plan with the bars of the E representing the two entrances and central detonation chamber; its walls are up to 61cm thick. A month later a more sophisticated detonation laboratory was designed for HER with a central reinforced concrete bursting chamber, divided into a small cell and a large cell (Fig. 3.25).[54] Explosive events in the chambers were observed through heavily armoured windows and recorded using state-of-the-art high-speed cameras. Experiments were monitored from an upper control room with a photographic darkroom below.

To the north of the Mobilisation Centre several standard Ministry of Supply huts were constructed to accommodate the office-based activities associated with electronics, engineering, mathematics and theoretical physics departments.[55] Also constructed in this area was a large store building. The most distinctive structure in this area is Q14 (Fig. 3.26). Unfortunately, no original drawings have been found for this building, but a member of the HER team recalled that it was colloquially known as the 'RAF building'.[56] This name reflects a notable feature of the development of the Mark 1 atomic bomb, namely the close cooperation between the intended service user, the RAF, and the HER team. Shortly after the decision was taken to proceed with the project in summer 1947, Squadron Leader John Rowlands was selected to head the RAF team that was to participate in the development of the bomb. His principal task was to ensure that the design of

Fig. 3.23
Fort Halstead, Kent. The eastern side of the Mobilisation Centre showing the buildings built for the High Explosives Research team, a scene little changed from the late 1940s. To the left is the bomb chamber F16, in the middle ground the recording laboratory F18 and to the left rear the detonation chamber F17.
[© Historic England Archive, DP060587]

Fig. 3.24
Fort Halstead, Kent. Bomb chamber F16 designed in August 1947, listed Grade II.*
[© Historic England Archive, DP060586]

Fig. 3.25
Fort Halstead, Kent. Detonation chamber F17 designed in 1949, listed Grade II.*
[© Historic England Archive, DP060573]

Fig. 3.26
Fort Halstead, Kent. Building Q14. To the right is its former main entrance and to the left the plant room marks the position of the original entry into the workshop. A plaque on the side of the building reads, 'In this building a group of scientists led by the then Dr W. G. Penney worked on United Kingdom atomic warheads during the period 1946 to 1952'. Listed Grade II.
[© Historic England Archive, DP060631]

Fig. 3.27
An intriguingly detailed full-size concrete model of the outer shell of the Blue Danube *'physics package' discovered at the former Royal Ordnance Factory, Elstow, Bedfordshire, in 2009. The circles mark the position of the initiators. It is about 1.5m in diameter and is on display at the Sywell Aviation Museum.*
[© Historic England Archive, DP436962]

the bomb followed the Air Ministry's Operational Requirement 1001 and that its components fitted in a casing compatible with the bomb bays of the projected V-Force aircraft.[57] The RAF team was also required to understand the science and engineering of the bomb, and to prepare the necessary procedures and regulations for storing, servicing and operating the weapon. It comprised Rowlands and a further nine serving RAF officers, all with strong backgrounds in mathematics, science and engineering. Once at Fort Halstead, the RAF contingent was closely integrated with the HER team. Rowlands headed one group that was responsible for weapon assembly, which also included Squadron Leaders Brown, Mitchell and Skelley, and Flight Lieutenant Blythe. Wing Commander Hunter-Toddy joined the mathematics team, Squadron Leaders Betts and Pulvermacher went to the electronics department, and Flight Lieutenant Mercer to explosives.[58] A general duties officer was also assigned to the team to look after operational matters.[59]

Q14 was a purpose-built structure designed for the assembly of the prototype warhead and its ballistic casing, and was central to Rowlands' task 'to see that everything we were making could be put together in one case'.[60] This building played a central role in the project and was the one place where all the highly classified components were brought together and assembled within the ballistic casing designed by the Royal Aircraft Establishment, Farnborough.[61] The only difference to an actual device was the substitution of inert wooden replicas in place of live explosive components (Fig. 3.27).

The architecture of Q14 directly reflects its specialised and secretive functions. It is a two-storey structure with a tall ground-floor workshop served by a travelling crane, and on the upper storey, offices. Originally, the eastern and southern elevations, facing out of the HER area, were blank and the workshop was lit by tall, and now partly infilled, windows on its north and west elevations. Entry into the building was restricted to two doors, one now blocked, on its southern elevation, with an internal staircase at this end of the building leading to the upper offices. On its west side was a single bay-wide entrance into the workshop. This was later blocked and obscured by a plant room, but a surviving wall light and metal covering, perhaps for a roller door, mark its position. The casing of the first-generation atomic bomb, codenamed Blue Danube, comprised three sections: a nose, central section and tail unit. The single-bay entrance was therefore wide enough to accommodate each section on a handling trolley, while the travelling crane would be required to move the entire bomb once it was assembled.

By the end of 1950, atomic weapons work dominated Fort Halstead's activities, with nearly 600 out of its allocated 997 posts allocated to HER. Atomic weapons-related work remained there until 1955, when the last members of staff were transferred to the new Atomic Weapons Research Establishment at Aldermaston (Berkshire).[62]

The Royal Arsenal, Woolwich

The Royal Arsenal, Woolwich, had been at the forefront of the development and manufacture of the most up-to-date armaments since the late 17th century. In 1950 just over 400 people were allocated to HER at Woolwich, and after Fort Halstead this represented the greatest concentration of project personnel.[63] The principal task of this group, led by Ernest Mott, was the development of the shaped blocks, or lenses, of conventional explosives that were required to initiate the nuclear explosion. This involved the manufacture of the separate lenses attached to metal backing plates held together with long piano-style hinges to create a hollow sphere of explosives that would surround the fissile core. The moulding, combining and machining of explosives to create the implosion wave represented a novel and potentially hazardous activity. At Woolwich, the team was able to draw on the expertise of the staff and research facilities accustomed to producing small experimental production runs of explosives. For testing, the lenses were transported for a relatively short distance across the Thames to the Shoeburyness range (Essex) where the project acquired a dedicated test range (*see* p. 60).

Fig. 3.28
Woolwich, Royal Arsenal East, September 1960. Showing the Research Department, the area occupied by HER was to the centre right. The two square buildings with black roofs were originally nitrating houses and the rectangular building to their left with a central court the Organic Laboratory.
[Historic England Archive (RAF Photography), raf_543_1059_1f43_0144 LN2078]

Fig. 3.29
Woolwich, Royal Arsenal East. The Organic Laboratory, constructed in 1939, was later incorporated into the HER section. Photographed in 1995, it has subsequently been demolished.
[© Crown copyright. Historic England Archive, BB95_00929]

Most of Mott's group was accommodated in the Research Department's Royal Arsenal East site, which comprised a handsome Neo-Georgian-style headquarters and various workshops. Immediately to its east a fenced enclave (Fig. 3.28) was created for work on the explosive lenses. This incorporated the Organic Laboratory completed in 1939 (Fig. 3.29) and to its east a small explosives section. At its eastern end were three part-finished and unoccupied explosives nitrating houses. One was demolished, and the other two converted, one into a radiological testing laboratory and the other into the radiological photographic building, which probably housed X-ray machines to detect minute cracks in the lenses that might affect their function.[64] The area eventually comprised about a dozen buildings, a mixture of modified structures and new-build. A reduction in security by late 1954 indicated that most activity had been transferred to Aldermaston by this date. This site has been cleared and lies beneath HM Prison Thameside.[65]

Atomic Weapons Research Establishment, Woolwich Common

Typical of many research establishments, Fort Halstead possessed light engineering facilities that might be used to produce prototype components but was not equipped to carry out small production runs and some types of specialist manufacturing. To provide dedicated light engineering support, HER acquired the former Radio Production Unit on Ha-Ha Road on the northern edge of Woolwich Common (Fig. 3.30).[66] This site offered a secure environment for the production of confidential components and was also able to produce one-off items, or to manufacture small production runs that were unattractive to private industry.

Another important factor in locating this work at Woolwich was the existing personnel's expertise, and the availability of unskilled and semi-skilled female labour for the assembly of electronic components.[67] Work carried out included the development and preproduction manufacture of electronic, electro-mechanical and light mechanical assemblies, instrumentation for trials work at home and abroad, and laboratory equipment. Many of these assemblies were crucial for the trials work on the fuzes and firing circuits for Britain's first atomic bomb Blue Danube, and it later produced components for the mid-1950s overseas trials codenamed Mosaic, Buffalo and Grapple.[68] William Penney later commented on the 'invaluable' work that had been carried out here, providing electronics and light engineering support for the early trials.[69] A small number of staff from the Explosives Inspection Department, known as P10(W), were also stationed on site and were solely responsible for the inspection of electronics and components associated with atomic devices.

In common with most government building programmes, at this date responsibility for the design and adaptation of the existing establishment lay with the Ministry of Works following specifications drawn up by the HER team. Work commenced on the site in 1950, with a complement of 150 people. In the following years it underwent further expansion, including the erection of steel-framed Marston sheds to house

Fig. 3.30
AWRE Woolwich Common, April 1967, showing the site shortly after it had been vacated by AWRE.
[© Historic England Archive, MAL3367]

the production of fibre-glass nose cones for bomb casings.[70] By the middle of the decade the workforce had risen to about 500, and in 1956 Penney proposed to expand it further to 1,450.[71]

Historic aerial photographs show that throughout the 1950s and early 1960s continued piecemeal development took place, including a canteen to serve the much-expanded workforce. Early buildings included standard Ministry of Works structures, such as concrete Orlit huts and Marston sheds, but in some of the later buildings, perhaps under the influence of the Atomic Weapons Research Establishment's architect's office, headed by A. Hancock, restrained decoration in the form of projecting brick headers was evident.[72] In common with

Fig. 3.31
AWRE Woolwich Common. A typical early 1960s research establishment office and laboratory block, designed under G. W. Dixon, UKAEA Southern Works Organisation.
[© Historic England Archive, DP093959]

most research establishments, the doors and other woodwork were latterly finished in light blue paint. One of the largest buildings was the 13-bay, two-storey office building constructed during the early 1960s on a reinforced concrete frame with buff brick infill panels (Fig. 3.31). This was designed by, or under, G. W. Dixon, architect at UKAEA's Southern Works organisation.[73] Facilities eventually included workshops for production machinery, sheet metal working, electronic assembly, woodwork, and plastics to produce plugs and sockets for high-voltage circuits. It also possessed extensive testing facilities, for temperature, pressure, vibration, high-voltage DC and AC circuits, and an X-ray machine. The establishment closed in 1965 and subsequent users retreated to the western end of the site before it was finally cleared for car parking space for the 2012 London Olympics.[74]

Atomic Weapons Research Establishment, Aldermaston

In common with other wartime establishments, the temporary airfield established at Aldermaston to the south-west of Reading in Berkshire offered basic infrastructure of use for atomic research, as well as being relatively close to London and Harwell (Fig. 3.32). The High Explosives Research Project moved its administrative headquarters from Fort Halstead to Aldermaston on 1 April 1950. Four years later it became the Weapons Group of UKAEA.

The immediate requirement was to construct the novel buildings needed to manufacture the plutonium core, uranium tamper, and other specialised components for the Hurricane atomic bomb trial scheduled for early autumn 1952. Most of the construction work was undertaken by Chivers and Sons of Devizes, which was also the main contractor at Harwell (*see* p. 88). To the north-west, the main processing area A, 'The Citadel', was constructed parallel to the runway and within it was the original plutonium and uranium processing and machining building A1-1 (Fig. 3.33). Here the plutonium initially produced by the chemical group at Harwell and later at Windscale was machined into a sphere, which could be unscrewed to insert the polonium initiator, or urchin. On completion the finished cores were gold-plated to prevent oxidation. Workshops in this area were also responsible for the uranium tamper, around which the explosive lenses were arranged.

The original A area buildings remained in operation into at least the late 1980s, although the discovery of contamination led to the closure of A1 between 1978 and 1982.[75] This probably hastened the decision to proceed with a new plutonium and uranium manufacturing facility, A90, with work commencing in 1983. The design of this large two-storey concrete structure closely followed that of the Plutonium Processing Facility, TA-55, at the Los Alamos National Laboratory (New Mexico), although modified for the local topography and climate. Internally there are around 300 glove boxes used to carry out a similar number of processes to create a warhead. It became operational in 1992 at a reputed cost of £1.5 billion.[76]

In addition to the specialised warhead production and assembly work, Aldermaston was, and remains, one of the country's best-equipped scientific research establishments. In the late 1950s around 10 per cent of its activities were concerned with basic physics.[77] At this time it hosted several research reactors (Table 3.1). HORACE, a light-water-moderated and -cooled reactor, was built to support general physics experiments, including how metals react under neutron bombardment. It was also used to refine the core arrangements for a future reactor HERALD, designed by AEI based on MERLIN, installed nearby at Aldermaston Court (Fig. 3.34). HORACE was housed in a 3.66m-diameter tank sunk 4.57m below ground, which later became the fuel store for HERALD, which ran until 1988.[78] In the late 1960s, VIPER (Versatile Intermediate Pulsed Energy Reactor) was installed to contribute to studies related to 'hardening' warheads against the effects of external radiation.[79] Other specialised experimental equipment included 3MeV and 6MeV electrostatic generators and a 800keV Cockcroft-Walton machine, used to generate very high voltages needed to produce highly energetic beams of charged particles.[80] In 1959, in collaboration with AERE and the Metropolitan-Vickers Electrical Company Limited, it also built a 12MeV particle accelerator contained in a pressure vessel 13.71m high and 2.74m internal diameter, located in a special building 32.92m tall.[81]

Mathematics was at the heart of research at Aldermaston, and in autumn 1953 a Ferranti Mark 1* computer was ordered, followed three years later by an English Electric DEUCE; both were valve machines and ran 24 hours a day, 7 days a week.[82] Prior to this, some AWRE

Fig. 3.32
Aldermaston, Berkshire, Atomic Weapons Research Establishment, 12 March 1954. This view shows how the 'A' pattern of wartime runways has influenced its layout. To the top right are the well-spaced structures of the explosives area, to the centre the regularly organised buildings of the plutonium and other metals machining area, below these buildings work is underway on an office, and at the bottom of the picture construction of the adjacent housing estate has begun. [Historic England Archive (RAF Photography), raf_82_866_f43_0175]

Fig. 3.33
Aldermaston, Berkshire. Building A1 in which the plutonium cores for the early nuclear weapons were prepared. [© Crown copyright. Photo courtesy of AWE, G3469_3]

Fig. 3.34
Aldermaston Court, Berkshire. Building R61 which housed the HERALD reactor. It is typical of many UKAEA buildings, having a concrete-framed office block with its exterior walls infilled with brick and patent glazing panels. The reactor building is steel framed clad in asbestos sheets, horizontal glazing and gravel-washed panels. [© Crown copyright. Photo courtesy of AWE, G7268_2]

Table 3.1 Aldermaston reactors and critical facilities.

Name	*Operational*	*Notes*
ATLAS	1952	Large critical assembly machine
ERIC	1952	Small critical assembly machine
WATERFALL	1954	Critical assembly machine
HORACE	1958	Swimming pool-type reactor U-235 fuel
HERALD	1960	Swimming pool-type reactor U-235 fuel
VERA	1961	Fast reactor studies
VIPER	1967	Versatile Intermediate Pulsed Energy Reactor

Sources: European American Committee on Reactor Physics 1964 *List of Research Reactors and Critical Facilities*; Hawkings 2000, 25–7.

computing had been undertaken at the National Physical Laboratory and the University of Manchester.[83] In the early 1960s, to match the computing capacity of the United States' weapons laboratories, an IBM 7030 Stretch computer was installed. This was the first transistorised super-computer, and at the time the world's fastest.[84] Despite growing American dominance in this field, in November 1963 AWRE issued a letter of intent to purchase a Ferranti/ International Computers and Tabulators Atlas 2, a slightly cheaper version of its Atlas 1 computer. This was formally accepted in 1966 and ran until 1971. Computer technology brought with it requirements for new building types. In 1956 a dedicated double-storey building was completed comprising ground-floor accommodation for punched card accounting machines with offices above. A year later the east computer room was completed to house an IBM 704 and in 1960 the west computer room was added for an IBM 7090. Its occupation was short, and it was replaced by the IBM 7030 Stretch computer.[85] Around 1970 a further wing was added to house an IBM 360 series machine. Distinctive features of this building included separate rooms to accommodate the Atlas, Stretch and IBM 360 computer cabinets, and control desks mounted on suspended flooring with cabling and air conditioning beneath. These were surrounded by plant and air-conditioning rooms, spaces for the engineers from the suppliers, magnetic tape store areas, a large punch card room, and other supporting facilities. In the beginning, scientific staff operated the computers, but by the early 1960s, to maximise the use of these expensive machines, they were joined by dedicated operators who took over during evenings, nights and weekends. The Atlas ran for 120 hours per week and the Stretch for 160 hours.[86] Contemporary photographs show that many of these operators were female. As with other facilities, the computers were shared with UKAEA colleagues at Culham, Harwell, Risley and Winfrith, in the 1960s absorbing about one-third of Aldermaston's computing time.[87] Further IBM machines were purchased until the late 1970s when again, following Los Alamos's lead, a Cray 1A was installed.[88]

Atomic Weapons Research Establishment, Burghfield, Berkshire

In 1954, in anticipation of a production run of around 100 warheads for the Blue Danube weapon series, Royal Ordnance Factory (ROF) No. 18, Burghfield was acquired for warhead assembly (Fig. 3.35). It is located about 8km east of Aldermaston and was a relatively new factory, completed in 1942 for the manufacture of 20mm Oerlikon cannon shells. In comparison to Aldermaston, it has an industrial character; its principal activities include the manufacture of specialised weapon components, warhead assembly, and the decommissioning of obsolete weapons. Visually the site retained its ROF layout into the 21st century, with a strong rectilinear plan, buildings surrounded by earthen safety bunds, masts for lightning protection, and runs of exposed, lagged, steam-heating pipes.

Renewed nuclear cooperation with the United States through the 1958 Mutual Defence Agreement, and the beginning of the manufacture of UK warheads influenced by US designs, led to new assembly buildings constructed in 1960. Known colloquially as the 'Gravel Gerties', they closely followed the form of structures at the Pantex assembly plant, Carson County

Fig. 3.35
Royal Ordnance Factory, Burghfield, Berkshire, 10 July 1946, before its acquisition by AWRE. To the top is the Grazeley Green camp used for post-war accommodation. [Historic England Archive (RAF Photography), raf_106g_uk_1646_rp_3202]

(Texas) (Fig. 3.36). To minimise the risk of flying debris, their concrete domed roofs were made of steel mesh sprayed with concrete. Around the exterior, voids filled with loose gravel were designed to smother any accidental explosion. As part of the proof of concept for this design, on Frenchman Flat, within the Nevada Test Site, USA, high explosives were detonated with depleted uranium in a prototype structure to test its effectiveness.[89] Its ruins, with masts for cameras and pressure sensors, remain, with signs warning of a possible radiation hazard.

AWRE Foulness, Essex

One of the many challenges of designing the bomb was ensuring that the detonators fitted to the explosive lenses worked reliably and simultaneously. Under the leadership of Ieuan Maddock, many hundreds were tested at Fort Halstead and at firing facilities at the Royal Arsenal.[90] But the project also required a trials range where multiple explosive lenses could be assembled and tested. Down river from Woolwich, and across the Thames, is the

Fig. 3.36
AWRE Burghfield, Berkshire, 2010. The white circles mark the roofs of the Gravel Gerties.
[© Historic England Archive, 26778_037]

Shoeburyness (Essex) firing range that had been active since the mid-19th century and remains so today. The ARD already had a presence on the range, and as part of its immediate post-war restructuring was intending to consolidate more of its activities on the still more isolated and remote Foulness Island further to the east.[91] Here many distinctive structures associated with nuclear weapons development survive, with some remaining in use.

As early as July 1945 plans were drawn up for the new 420ha Shelford Range on Foulness Island, and in summer 1947 it was allocated to HER.[92] Foulness was to be a purpose-built, permanent establishment, and this is reflected in the robust quality of many of the buildings. Most of the range buildings are single-storey, brick-built, with flat, asphalt-covered concrete roofs, lit by metal Crittall windows protected by steel shutters. Curiously, little use was made of reinforced concrete, even for the range control buildings, which may reflect post-war shortages of building materials and ready access to the products of local brick yards. Notwithstanding the high priority of the project, in November 1947, William Penney was forced to write to Lord Portal, Controller of Atomic Production at the Ministry of Supply, complaining about a possible 12-month delay to the programme caused by construction holdups.[93]

For safety, but also security, reasons, the range was divided into several discrete functional areas. At the western end of the range, over 2km from the main gate, was the administration area. Initially this area comprised 18 buildings including a security lodge, search room, offices, a photographic laboratory, stores, workshops, and a canteen (Fig. 3.37). The largest building in the area is the double-storey headquarters, reflecting the permanent nature of this site. In contrast with the utilitarian quality of the remainder of the buildings, above its door a stone plaque records 'A.R.E. 1949', which was lit by an ornate lantern.

One of the main functions of the Foulness range was to provide space to test fire the explosives lenses that surrounded the bomb's core. As explosives arrived on site, including the experimental lenses produced at the Royal Arsenal, they were taken for storage and preparation to

Fig. 3.37
AWRE Foulness, Essex. A photograph of the headquarters area taken during the assessment of the devastating flood of February 1953, showing the area just after Britain's first atomic bomb test and before the start of the hydrogen bomb programme.
[Historic England Archive (RAF Photography), raf_82_708_0190]

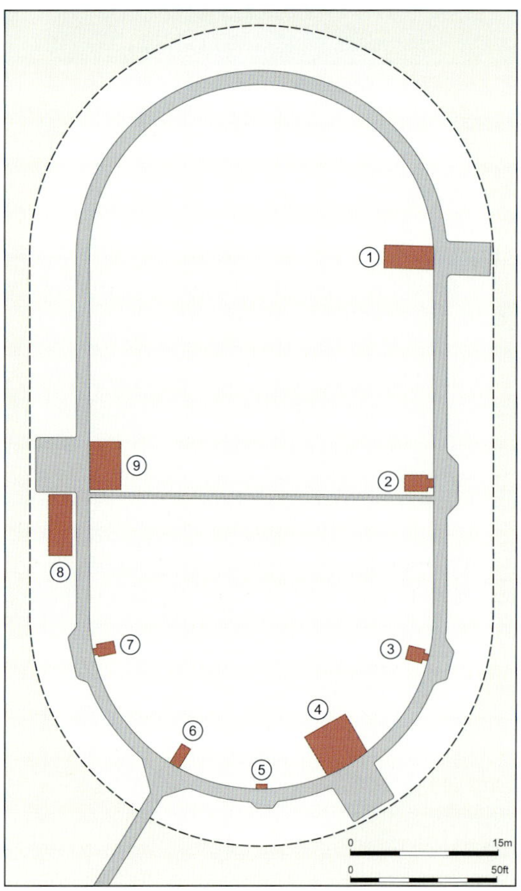

the Explosives Storage Area at the northern end of the range. This was constructed adjacent to Smallgains Farm and a pair of First World War, War Department, cottages that remained in residential use by site personnel. The elliptical layout of the Explosives Storage Area was probably inherited from the initial post-war plan, but several building designs were modified to meet the specific needs of HER. Entry into the area was through a single gate controlled by the Police Post and Search Room (Fig. 3.38).

Central to the development of the lenses was Building 23, the Explosives Preparation Laboratories (Figs 3.39–3.41). The term 'laboratory' in ordnance parlance described a building where loose explosives were handled and assembled. In a drawing signed off in December 1947 by the range superintendent, Roy Pilgrim, one of the British observers on the United States Bikini Atoll trials, its size was to be more than quadrupled from 214 cu m to 990 cu m.[94] Its brick construction is similar to other range structures. Entry is through a single copper-clad door that gives access to a small shifting lobby where special magazine-type clothing was put on, including metal-free shoes, every effort being made to eliminate the risk of sparks.

From the changing area personnel could enter the office or one of two smaller laboratories. Most recently, Laboratory 3 was used as a metrology room, with a large and heavy steel table at its centre for the precise measurement of components. The scale of Laboratory 1 is consistent with the handling of the 1.52m-diameter sphere that formed the central section of the first generation of British atomic bombs. Material for assembly was brought in through a pair of

Fig. 3.38
AWRE Foulness, Essex. Plan of the explosives area, 19 December 1947.
[Redrawn from TNA ES1/330]

Key
1 Bomb and shell store
2 Bulk high explosive and bare charge store
3 Powder store
4 Explosives casting and pressing laboratories
5 Detonator store
6 Store for munitions which may not be stored in other explosives stores, Building 22
7 Fuze and flash store
8 Non-explosive components store
9 Explosives Preparation Laboratories, Building 23

Fig. 3.39
AWRE Foulness, Essex. Explosives Preparation Laboratories, Building 23. During summer 1952, the country's first atomic devices were assembled in this building. It is a scheduled monument. [© Historic England Archive, DP035925]

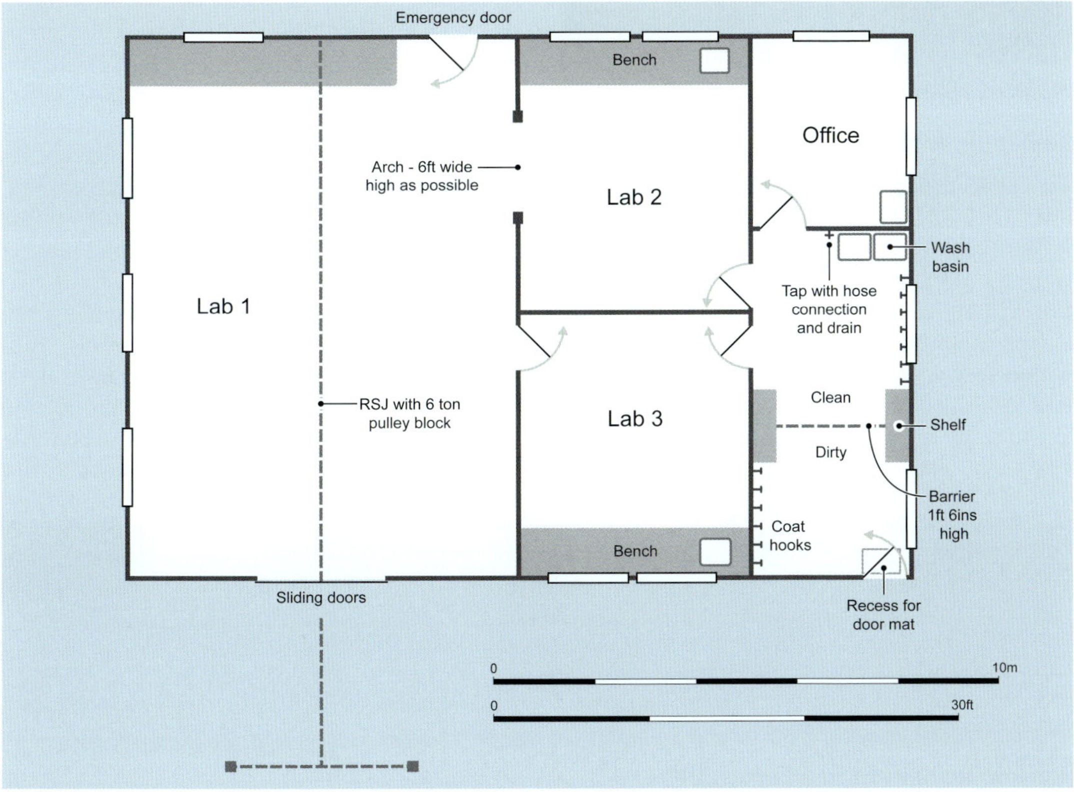

Fig. 3.40
AWRE Foulness, Essex. Explosives Preparation Laboratories, Building 23, redrawn from a plan dated 18 November 1947. [Redrawn from TNA ES1/330]

sliding doors, utilising a rolled steel joist with a 6.1-tonne capacity, which was more than adequate to lift an assembled device. This opened into Laboratory 2; its entrance is 1.83m wide, which may indicate that some assembly work was also carried on in this room. Building 23 provides a good example of the maxim that 'form follows function' and there is a clear resemblance to a building on V-site, Los Alamos (New Mexico), where the first implosion weapons were assembled.[95]

Other buildings supporting the activities in this area included the bomb and shell store, powder store, bulk HE and bare charge store, detonator store, fuze and flash store, a store for munitions which may not be stored in other explosives stores (Fig. 3.42), explosives preparation laboratory, and the non-explosives components store.[96] The use of copper-clad doors on some of these buildings gives an archaic appearance reminiscent of much earlier black-powder magazines.

Fig. 3.41
AWRE Foulness, Essex. Explosives Preparation Laboratories, Building 23, Laboratory 1. The main assembly area, showing the modified entrance, door to Laboratory 3, overhead crane beam, and sealed safety lighting. It is a scheduled monument.
[© Historic England Archive, AA99_09785]

Testing the lenses

The novel technology of using lenses formed from high and low explosives to focus a blast wave on a fissile core was developed during the wartime Manhattan Project and was known to the British scientists involved in the work. Principal among these was William Penney, whose speciality, hydrodynamics, was crucial to understanding the behaviours of blast waves. Technically, the manufacture of the lenses was extremely challenging, as they had to be produced to very fine tolerances and free from any internal cracking that might develop during cooling. The precision of this assembly work was critical to its operation, as any misalignment of the lenses might result in the misdirection of the explosive shock wave onto the fissile core. Initially, a few lenses were combined, and in April 1949 a six-unit trial was staged – the whole bomb comprising separate hexagons and pentagons of high explosive.[97] Later, a hemisphere, representing about 2.54 tonnes of explosives, was assembled and fired, some of these detonations resulting in claims for smashed windows in Southend-on-Sea, 10km to the west.[98]

The ranges

From the Explosives Preparation and Storage Area, assemblies were taken to one of four firing ranges laid out to the south of the main site access road (Fig. 3.43). At the eastern end of the range a farm complex was demolished to make way for Range 1, or the Large Bomb Range. At the range head, close to the main access road, were an office and electronics laboratory, toilet, and mess room. Control and instrumentation for tests was originally carried out from two recording buildings, one to the east and one to the west, both situated over 100m away from the firing sites. In the early 1950s these were supplemented by two large reinforced concrete structures, a photograph recording a building with 11 square steel shutters, and a control and

Fig. 3.42
AWRE Foulness, Essex. Building 22, special store. Built in 1947, it is typical of this date being brick built with a flat concrete roof and copper-clad doors. Characteristics of an explosives handling building include the lightning conductors, copper earth straps and external electrical fittings. It is a scheduled monument.
[© Historic England Archive, DP035923]

Fig. 3.43
AWRE Foulness, Essex, 30 June 1955. To the left is Range 1, or the large bomb range, with two scars caused by detonating large amounts of explosives. Some of the faint straight lines mark the position of control and command cables running from the protected buildings. [Historic England Archive (RAF Photography), raf_542_201_0102]

instrumentation building protected on its range side by 25mm steel plate.

The Hurricane trial

In summer 1952, in the lead up to the autumn Hurricane trial, in which Britain would detonate her first atomic test device in the antipodes, Building 23 was the main focus of activity. Here, Ernest Mott, and George Gallie from Woolwich, led the assembly team, which included H. S. Weeks and a group of technicians, including Mr Hessen who was in charge of metrology.[99] Initially they practiced on an inert model, code-named Alfred, followed by the assembly of up to three live devices, Hero, Hengist and Horsa. The latter two were named after two semi-legendary early 5th-century Jutish warrior brothers, who along with their war band were invited to defend Kent from the Germanic tribes.[100]

The range at Foulness was also used to develop the monitoring instruments and to train the personnel who would operate them during the overseas tests. Preparations included the detonation of a 3,629kg charge to test photographic equipment and the coordination of the observation teams. The signal to detonate the device was to be sent by cable, while data from the various observation points would be carried back to the control centre by state-of-the-art radio equipment. To practice the test routines, firing signals were sent about 10km from Shoeburyness to the war-surplus river-class frigate HMS *Plym*, the vessel identified to carry the test device, moored at Stangate Creek (Kent). As the date of the test approached, more elaborate tests were conducted to synchronise the firing signal with the monitoring equipment, some involving the firing of a 29kg charge at Foulness.[101]

On Thursday 5 June 1952, one or more of the devices were taken by lorry to Shoeburyness and then by barge to HMS *Plym*. From here, she was escorted by the trials flagship HMS *Campania*, just released from her Festival of Britain duties, for her eight-week voyage to the Monte Bello Islands test site (Fig. 3.44). After their arrival, a further eight weeks were spent erecting the structures and equipment that were to be subjected to the full force of the device and preparing the firing circuits and monitoring equipment. The fissile core, manufactured at Aldermaston, was flown to Australia by Sunderland flying boat – it arrived on 18 September. A device was successfully detonated on Friday 3 October 1952 producing an estimated yield of 25 kilotons, 4 kilotons greater than the similar bomb dropped on Nagasaki in August 1945.[102]

Fig. 3.44
HMS Campania, *during 1951. As the Festival of Britain ship, she toured the country, and a year later she was the headquarters ship for the* Hurricane *trials.*
[Author's collection]

The hydrogen bomb

Following the Cabinet decision on Monday 26 July 1954 to proceed with the development of the hydrogen bomb, or H-bomb, staff numbers at Foulness were predicted to grow from 297 to 408 by March 1955, about a quarter of whom were scientific staff.[103] In addition to work on the H-bomb, development was also continuing on smaller tactical nuclear weapons, including Red Beard and the later abandoned Pixie warhead, in addition to the refining of trials instrumentation. Ernest Mott, newly appointed as superintendent, drew up a plan for the rapid enlargement of Foulness, in his words, 'in anticipation of a considerably increased and broadened programme'.[104]

To accommodate the increased staff numbers and new research programmes, there was a marked expansion in the facilities. The headquarters area tripled in size, and the urgency of the work was reflected in the use of a contemporary prefabricated lightweight building system.[105] This combined economy of materials with speed of erection, while presenting a modern industrial design (Figs 3.45 and 3.46).[106] The main elements of the system are internal frames of hollow-steel square section members. The roofs are supported on lightweight flat steel lattice trusses, and internally are insulated with compressed wood straw panels. Externally, the buildings are faced in uniform washed cement panels, although on the upper sections of the workshops pressed stainless steel sheets were also used. They are glazed by standard steel-framed Crittall-type windows. The pedestrian doors to the buildings are wooden with concrete or pressed metal canopies supported on tubular steel columns, while the larger openings to the workshops and stores are entered through sliding doors. Internally, they are either lined in brick or had a cement rendered surface and are often painted pale blue. This system was used to provide workshops, laboratories, stores and a two-storey office building.

After this brief flirtation with this lightweight building system, later buildings returned to sturdier brick-built forms or used concrete frames with brick infill (Fig. 3.47). The enlarged staff was also reflected in the need for a larger two-storey plain brick-built administration building. The late 1940s welfare facilities were also no longer adequate for the expanded workforce, and in 1956 a new canteen was added (Figs 3.48 and 3.49). Utilitarian in appearance, it is a light and airy building constructed in a contemporary style. It is well finished, with its roof supported on chamfered concrete columns, with large, glazed elevations on all sides except to the north where there is a lower aisle that housed the kitchen. The large windows gave views across the marshland, and to the south the large grassy area between it and the perimeter fence was often used for impromptu lunchtime football matches. Its architecture clearly illustrates the

Fig. 3.45
AWRE Foulness, Essex, main mechanical workshop E2, 1955. Its entrance exhibits many features typical of this building system, including hung concrete panel walls, steel-framed windows, pressed aluminium cladding, and a porch supported on a tubular steel framework. [© Historic England Archive, DP035959]

Fig. 3.46
AWRE Foulness, Essex, main mechanical workshop E2, showing its steel framework construction and additional framing for an overhead crane. The lightweight roof trusses are formed from steel sections with tubular steel struts. [© Historic England Archive, DP035920]

rigid civil service and industrial hierarchies of the 1950s. The main entrance is on the eastern side where three of the bays are inset to create a covered porch with two pairs of doors to the rear. They lead to a large open dining room floored with a chequerboard pattern of blue and white linoleum tiles and lit by pendant lights with large white glass shades. Adjacent to the kitchens was a servery, and to the rear a carpeted bar area separated from the dining room by a partition with copper panels decorated with embossed abstract designs. On the west wall was a large world map. At the rear of the building a separate entrance for senior staff gave access to a small lobby area with toilets, and then into a smaller dining area with wooden parquet flooring.

Test bans

During the 1950s there was growing worldwide concern about increasing amounts of radioactive debris entering the atmosphere through weapons tests.[107] In addition to other political factors, this threat of a suspension of atmospheric

Fig. 3.47
AWRE Foulness, Essex, showing the system-built Laboratory Ll, 1955, and to its right is the slightly later traditionally brick-built electronics laboratory L6. [© Historic England Archive, DP035967]

Fig. 3.48
AWRE Foulness, Essex, canteen A1, built in 1956, showing the main entrance. [© Historic England Archive, DP035913]

Fig. 3.49
AWRE Foulness, Essex, canteen A1. Main dining area with its linoleum-covered floor and beyond, the bar area, with a world map to the rear. [© Historic England Archive, DP035914]

testing acted as a compelling impetus for Britain to prove her weapons designs as quickly as possible. A moratorium on nuclear tests was finally agreed between the United States, Soviet Union and Britain in October 1958.[108] This lasted until the Soviet Union renewed atmospheric testing in September 1961. Ultimately, the Partial Test Ban Treaty, signed on 5 August 1963, brought an end to all atmospheric nuclear tests, and in outer space and underwater. This left open the possibility of underground tests; one method to verify and gather intelligence on such trials is through forensic seismology by locating and measuring the strength of seismic waves created. However, Dr Albert Latter of the American RAND Corporation hypothesised that the seismic signature of an underground explosion might be decoupled by suspending a device within an underground cavity. To validate this theory, under the codename Operation Orpheus, AWRE carried out experiments in two mines. Small preliminary tests were carried out in the abandoned Excelsior Tunnel mine at Callington (Cornwall), while by November 1959 access had been secured to the larger Greenside lead mine (Westmorland and Furness). Experiments began here later in the year and ran on until April 1960.[109] A specialist geophysics section was also established close to Aldermaston at Blacknest, which remains active.

Blast simulation

These restrictions on nuclear testing brought new challenges for AWRE's scientists and infrastructure designers. It ended access to test data from atmospheric explosions, which might be used for developing and verifying warhead designs, stockpile stewardship, and for the investigation of the effects of nuclear explosions on biological specimens, structures and military equipment. For modelling purposes, the results of a nuclear explosion may be divided into four main effects: blast, heat, electromagnetic pulse and radiation. One method used to study blast effects is by detonating quantities of conventional high explosives, which then might be mathematically scaled to represent larger explosions. These trials might be used in combination with scale models to study the effects of blast on different structural types. From the early 1950s shock tubes were used to study blast effects in a more controlled environment. The smaller blast tubes relied on a build-up of compressed air that was contained by a diaphragm of known strength before releasing a shock wave down the tube.

With the loss of potential data from atmospheric tests in the mid-1960s, a large 201m blast tunnel was constructed at Foulness (Fig. 3.50). In this facility a far larger explosive wave was originally produced by blank charges fired

Fig. 3.50
AWRE Foulness, Essex, blast tunnel, showing to the left a movable section on which test items were placed.
[© Historic England Archive, BB99_15990]

by two naval guns. This method was later replaced by explosive mining cord. The resulting shock-wave was propagated down the tunnel where its effects could be measured on various type of military equipment, including tanks and complete airframes. Close by, to simulate heat effects, a Thermal Radiation Facility was added, fuelled by kerosene, liquid oxygen and aluminium powder. At the same time, electro-magnetic pulse and radiation effects facilities were built at Aldermaston. Ultimately, the designers needed access to actual warhead performance data, and this was provided by joint underground tests conducted on the United States' Nevada Test Site.

Later history

From the high point of the late 1950s, the prospect of sustained nuclear weapons research diminished during the 1960s as defence cutbacks took a toll on high-tech weapons projects, and the utility of certain types of warheads began to be questioned. Access to American nuclear know-how also began to reduce the need for British research, although investigations continued to anglicise American designs to meet United Kingdom safety and engineering standards. Furthermore, as warhead technology matured, it was possible to standardise on a small number of variants, some of which remained in service for several decades. As financial constraints cut into defence programmes, few new buildings were erected after the 1950s. The focus of work also changed, and the range also became involved in programmes to verify the safety of civil reactor containment vessels.

Despite the lack of new building, many of the existing buildings were continually modified, and new and updated monitoring equipment installed. To configure the firing areas for different trials they were also subject to repeated modification. This might involve the erection of steel frameworks or, more commonly, the creation of one-off structures using movable concrete blocks, known as Pendine blocks, after a range in South Wales. In addition, various portable trials cabins and armoured observation posts were positioned as required.

Atomic Weapons Research Establishment, Orford Ness, Suffolk

To date, the nuclear devices that had been created were scientific objects, produced and handled by scientists, engineers and specialist technicians. Before they could be issued to the services, they needed to be made more robust, safe to handle, and put through rigorous testing; in modern jargon they needed to be 'weaponised'. In 1954, a requirement was identified for facilities to simulate the vibration and changes in temperature differences a bomb might experience during flight. The immediate need was to provide facilities for a new smaller tactical weapon, Red Beard, and to test the Mark 1 atomic bomb before the Buffalo trials at Maralinga, Australia, set for October 1956.[110]

At first it was proposed to site these facilities at Foulness, but due to insufficient space to ensure safety distances between structures, and uncertainties about ground loading, this idea was abandoned.[111] William Penney was increasingly concerned about the delay in the provision of test facilities and how this might affect the trials programme, and wrote to the Ministry of Supply in March 1955 pressing for these facilities by spring 1956.[112] Probably prompted by Penney's letter, the RAE range at Orford Ness was identified for the new establishment, and work was underway by summer 1955 (Fig. 3.51).

Orford Ness, on England's east coast, to the south of Aldeburgh (Suffolk), is one of the largest vegetated spits in Europe and for that reason it was acquired by the National Trust in the 1990s, then primarily for its natural history interest. The spit was initially developed by the War Office during the First World War, when an airfield was constructed, mainly devoted to experimental activities. These continued through the inter-war period; most notable was work by Robert Watson-Watt's team to prove the technique of radio direction finding, or radar. Facilities were also added to study bomb ballistics. During the Second World War, the establishment was chiefly concerned with studying the effects of different types of weapons on airframes.

The association of Orford Ness with the atomic bomb project began in the late 1940s with investigations into the ballistic properties of the Blue Danube casing, as well as testing release mechanisms and its complex internal electronics.

Fig. 3.51
AWRE Orford Ness, Suffolk. By July 1958 the first phase of development was complete. To the right is Laboratory 1 and to the left Laboratory 2. Laboratory 3 is between them, its mound merging with the disturbed shingle.
[Historic England Archive (RAF Photography), raf_58_2519_0001]

The late 1950s

The function of the new AWRE facilities was characterised by a later document as 'adequate to cope with the needs of the Weapons Group to test devices under special handling conditions of overseas nuclear tests'.[113] Several Royal Flying Corps' buildings were refurbished, support buildings added, and a new causeway was built to move building material onto the spit. Initially, three laboratories were built. The first generation of air-dropped atomic bombs, codenamed Blue Danube, were large weighty weapons, which measured 7.3m in length with a maximum diameter of 1.52m, and when fully assembled weighed 4,636kg. The new test structures needed to be large enough to accommodate these, as well as robust enough to resist the accidental detonation of around 2.54 tonnes of conventional explosives that surrounded the fissile core.[114] Acting on behalf of UKAEA, these structures were designed by C. W. Glover and Partners, Consulting Engineers and Architects, London, at a projected cost of £750,000.

At the eastern end of the site is the Large Vibration Laboratory, Laboratory 1 (Fig. 3.52). It is a massive reinforced concrete structure with a large central bay measuring 30.69m by 9.13m. Internally, the main bay is separated into two unequal portions by a substantial concrete dividing wall, and running along its northern side is a pit 3.18m wide and 2.74m deep. This was used to house the cradle on which the bomb sat, and associated test machinery. In case of an accidental explosion, the sunken bay would offer some protection. To further guard against the effects of an accidental explosion, its concrete walls were given extra mass by an external shingle traverse, which would force blast vertically rather than laterally. The roof was a relatively weak steel-framed structure, clad in aluminium

Fig. 3 .52
Orford Ness, Suffolk, Laboratory 1. To the left is a detached plant room. It is a scheduled monument. [© Historic England Archive, DP068447]

sheeting, insulated with cork and covered in weatherproof felt. As the temperature and humidity control within the testing chamber was critical, the separate entrance lobbies to each bay served as atmospheric controls: each measured 10.1m by 9.2m and was sealed with wooden doors at either end. Although the original specification only called for the testing of the bomb's central section, the lobbies were long enough to accommodate a fully assembled Blue Danube and its transport trolley. A ledge below the top of the longitudinal walls of the central bay marks the position of a travelling crane that was used to position the bomb on the test equipment. Surrounding the main chamber, other rooms accommodated switch and monitoring equipment and staff facilities. On the west side of the mound is a large self-contained reinforced concrete, air-conditioning plant room which maintained the atmospheric conditions in the test chambers. On the north side of the traverse is a similar self-contained structure; surviving fittings show that it mainly housed electrical switch gear, controls for the vibrating machines and carbon dioxide supplies for the fire suppressant system.

Fig. 3.53
Orford Ness, Suffolk, Laboratory 1, interior. To the right the pit housed the main vibration testing equipment and beyond the opening was the drop-test facility. It is a scheduled monument. [© Historic England Archive, AA021787]

This building was designed to undertake two principal functions. The larger bay was the Main Vibration Laboratory, and its role was to subject an assembled weapon to sustained vibration to ensure that it was able to survive the rigours of flight and operate as designed (Fig. 3.53). The smaller bay was the Drop Test Laboratory. This was designed to carry out a standard military acceptance criterion that requires a weapon to be robust enough to withstand a handling mishap without detonation. These operations were so hazardous that it was provided with a detached control room 130m to its north-east. This is an L-shaped reinforced concrete structure, originally with an open-sided porch on its north-east

Fig. 3.54
AWRE Orford Ness, Suffolk, Laboratory 2, the probably secondary centrifuge pit. The ledge high on the walls formerly supported the rails of a travelling crane. [© Historic England Archive, AA021774]

side. This comprised two elements; a taller section to the north-east that housed air-conditioning plant, an entrance passage and toilet cubicles, and to its south-west was a large open room with a suspended floor that housed the monitoring equipment. The side of the building facing Laboratory 1 is protected by a shingle traverse.

Laboratory 2 is sited at the western end of the site and is of similar construction to Laboratory 1 (Fig. 3.54). It has a single large central bay 30.67m by 9.05m with projecting ledges on its walls that mark the position of an overhead crane. To either side are aisles running the full length of the central chamber. These have a distinctive roof structure, formed from concrete beams with curved breeze blocks between them; above they are covered by a shingle mound. At the western end of the main chamber is an entrance passage with double doors to either end, with enough space between them to accommodate a fully assembled Blue Danube on a trolley. There was a similar entry to the west that gave access into the northern aisle, with two smaller openings into the main chamber. To the west of the main building is a separate plant room. The original function of this building is unknown, but it may have been used for the assembly, storage and investigation of test rounds. If this was the case, the different sections may have been brought into the western aisle and moved into the main bay for assembly before being moved to Laboratory 1 or 3 for testing.

Between these two laboratories is Laboratory 3, which was used for climatic testing, the simulation of temperature variations a weapon might be subject to during service. The entrance to this building is protected by a large reinforced concrete canopy; beneath this was a steel runner beam, which ran into the building and was used with a trolley hoist to manoeuvre test items into place (Fig. 3.55). The interior comprises a large central chamber, 24.5m by 6m, which is divided into two sections (Fig 3.56). To the front was the preparation area and to the rear was an insulated chamber, 13.5m in length, which was separated from the preparation area by a cement-covered 'expandomesh' wall and a metal-clad sliding door carried on an overhead rail. The working area of the rear chamber measured 9.14m by 3m by 3m. The temperature range that could be achieved extended between +60 and −60 degrees Celsius, and while

Fig. 3.55
AWRE Orford Ness, Suffolk, Laboratory 3, climatic test chamber. To the left are various plant rooms and to the right the entrance porch, which also provided blast protection. It is a scheduled monument.
[© Historic England Archive, AA021783]

Fig. 3.56
AWRE Orford Ness, Suffolk, Laboratory 3, climatic test chamber. To the rear is the test cell and in the foreground, the displaced door.
[© Historic England Archive, DP068470]

undergoing thermal tests an object might also be subject to vibration.[115] This laboratory was also self-sufficient and had large plant rooms set into its west side.

The 1960s

In 1959, the requirement for new facilities to investigate weapons in service use was identified. Was it better to abandon the inconveniently located Orford Ness range and look for somewhere closer to Aldermaston, or to persevere with a place with heavy recent investment? After an extensive search for sites offering equivalent safety margins and security, it was concluded, less than enthusiastically, in January1960 'that there was no satisfactory alternative to continuing at Orford Ness'.[116] In the event, in common with other AWRE sites, the prospect of sustained research work diminished as defence cutbacks took a toll on high-tech weapons projects.

Later structures illustrate the range of work undertaken by the Southern Works Organisation of UKAEA, under their chief architect G. W. Dixon, to design and construct structures to the exacting specifications of the scientists. Detailed structural drawings were again prepared by C. W. Glover and Partners.[117] Construction was undertaken by the local builders and contractors Cubbitt and Gotts.

The Pagodas

The first facilities at Orford Ness had been closely modelled around the requirements of handling the Blue Danube bomb casing and its large conventional explosive component, but this casing was becoming obsolete, and new weapons and test programmes required bespoke structures.

Once the decision had been made to remain at Orford Ness, work proceeded quickly on two vibration test buildings, commonly referred to as the Pagodas, their control room, a magazine, and associated infrastructure, at an approved cost of £860,000 (Fig. 3.57). To reduce interference with ongoing test activities, a new track was built from the quay and another bridge was constructed over the Stony Ditch watercourse.

The specification for the Pagodas included an ability to withstand the accidental detonation of 181.4kg of high explosives, requiring a design which was quite evidently 'beyond normal architectural or structural practice'.[118] Experiments were carried out at AWRE Foulness using small amounts of explosives and one-tenth scale models. The initial proposal was for a monolithic structure with a single entrance; a detonation in a test piece destroyed the model and scattered large fragments. In the final version, the roof was given greater mass by piling gravel on top; it also overhung the test cell to contain the danger of flying debris being ejected from the building. When this form was tested, the roof and mound rose up, with part of the force being vented through the gap between the top of the cell and the roof; the columns then gave way allowing

Fig. 3.57
AWRE Orford Ness, Suffolk, the vibration test laboratories, or 'Pagodas', two of the most emblematic Cold War structures in England. They are scheduled monuments. [© Historic England Archive, DP068510]

Fig. 3.58
AWRE Orford Ness, Suffolk. The main test cell of one of the Pagodas with slots for mounting test and monitoring equipment. [© Historic England Archive, DP068518]

the roof to fall back into the cell.[119] Interestingly, a similar design had been used in the early 1950s at the Explosives Research and Development Establishment, Waltham Abbey (Essex), but perhaps reflecting the secretive nature of this establishment no reference was made to these structures.[120] Construction of the two vibration test buildings, at a cost of £120,000 per structure, was practically complete by June 1960. Work on the magazine probably did not begin until a year later when approval was given for the expenditure of £29,000.[121]

The vibration test buildings are nearly identical and comprise a large reinforced concrete central cell 16.47m by 7.30m covered by a massive reinforced concrete roof supported on 16 reinforced concrete columns; on two sides are self-contained plant rooms. There is a single main entrance into each building which gave access onto a lift platform that was used to lower test pieces into the central chamber, which was below ground level to further contain any accidental explosions. In the entrance passage a door provided entry to the main test chamber and an upper walkway.

To assist in manoeuvring heavy objects there are several substantial steel eyelets secured into the underside of the roof. A travelling crane also ran on rails mounted on a ledge beneath the windows. The floor of the main cell is formed of parallel and narrowly spaced steel I beams, which allowed test rigs to be firmly secured to the structure (Fig. 3.58). Slotted steel plates in the walls of the cell also enabled test and monitoring equipment to be fixed to the structure. In addition to vibration tests, objects might also be placed in jackets to simulate extremes of heat and cold, or in a portable pressure chamber to mimic the effects of altitudinal changes.[122]

Trials in the Pagodas were controlled and monitored from the control room about 130m to their north (Fig. 3.59). This was constructed in a simple, functional and contemporary style in Burwell white bricks that contrasted with the red asphalt-covered lower sections of the metal-framed bay panels and yellow-tinted Perspex windows. Internally, it is divided into two parts. The slightly taller section housed control and monitoring equipment, while the lower section contained various staff facilities. It is entered through an inset covered porch, which is surfaced in red and yellow paving slabs arranged in a chequerboard design. The taller western section of the building is split into two by a longitudinal corridor. Either side of this are six almost identical bays, with two forming a single large room. The control and monitoring equipment in these bays was supported on a raised floor; the under-floor void housed air-conditioning pipes and cabling. To reduce noise, the upper sections of these rooms and their ceilings were lined with square, pierced, sound insulation panels. Close to this building were a substation, pump house and water tanks.

It was during this phase that a magazine was built to the north, which perhaps replaced Laboratory 2 for the storage of test rounds. It is a reinforced concrete structure with barrel-vaulted roofs, revetted on all sides by shingle traverses. The entrance is a drive-through lateral passageway, which had a steel runner beam for a travelling hoist above. Along one side are two self-contained storage magazines, originally separated from the passage by toeboards denoting an inner clean magazine area from the outer dirty area. The two bays differ slightly in size, one measures 7.45m by 4.55m, and the other 9.15m by 7.45m; both are 5.12m high. Each was ventilated through a roof vent with a tall vertical stack projecting through the roof and by a pair of vents served by a rear air-conditioning plant room.

Centrifuges

In the early 1960s modifications were made to Laboratory 2 with the installation of a Napier centrifuge for testing components under extreme G forces (*see* Fig. 3.54).[123] This was housed in a brick-built circular enclosure 9.14m in diameter. The centrifuge arm was 4.88m in diameter and could support a test piece weighing up to 453.6kg.[124] A raised control room was probably also built at this time and the plant room was also modified.

In about 1966, at the western extremity of the site, work began on a new centrifuge. The design of the building suggests that it too was designed to handle devices containing explosives. It is a single-storey rectangular concrete structure with the two sides facing other buildings protected by shingle traverses. Internally is a sunken brick-lined centrifuge pit, which is 8.06m in diameter and 2.68m deep. Along the top of each of the long walls is a ledge that supported the rails of a now-removed travelling crane, which was used to install and service the centrifuge and lift test pieces into place. It is covered by light, pressed-metal roof sheets.

Fig. 3.59
AWRE Orford Ness, Suffolk, control room. It is a scheduled monument.
[© Historic England Archive, DP070009]

Fig. 3.60
AWRE Orford Ness, Suffolk. The development of the establishment to April 1965. To the top is the site of the Royal Flying Corps aerodrome.
[© Crown copyright. Ordnance Survey os_65054_V_108]

Significance of Orford Ness

The landscape created at Orford Ness in the late 1950s embodied world-class technology and high politics (Fig. 3.60). It was one part of a complex scientific and industrial infrastructure that had been built to supply the country's armed forces with nuclear warheads. Through adhering to tight timetables, the scientists, engineers and technicians crucially contributed to maintaining Britain's international political status. The detonation of an air-dropped atomic bomb in October 1956, followed in the next year by a thermonuclear device dropped off Christmas Island during the Grapple trials, demonstrated, not least to the United States, the country's scientific and technical competence. It was in part these events that led to the signing of the 1958 Mutual Defence Agreement, establishing among other provisions renewed British access to American nuclear technology.[125]

By the late 1960s, there was relatively little new work in prospect, and in April 1969 the decision to close Orford Ness and to move many of its functions to the Environmental Test Laboratory at Aldermaston was announced. The last trial took place on 9 June 1971 and the establishment closed on Friday 1 October 1971.[126]

A national effort

Historically, the interlocking networks of educational, technical, industrial expertise and infrastructure needed to support complex technological programmes have rarely been studied.[127] Beyond the government research establishments, the expertise of private industry was vital to the success of the atomic bomb project. Some of the diverse firms that contributed to the project included Hudswell Clarke and Co. Ltd, in the former Blackburn Aircraft factory, Roundhay Road, Leeds, which was entrusted with the manufacture of the bomb's casings.[128] The complex interlocking outer plates surrounding the sphere of explosives and fissile core were produced by Percival Aircraft, in a now-demolished factory on the site of Luton airport.[129] A most unlikely contributor to the project was William Freeman & Company Limited, Subaseal Works, Peel Street, Barnsley (West Riding of Yorkshire). Among their usual products were hot water bottles. Adapting this manufacturing know-how, they produced the rubber bags required to keep the explosive lenses in the metal casing under gentle pressure.[130]

Conclusion

Driven by Cold War political imperatives to develop a British atomic bomb, scientists, engineers and other technical staff working in government, universities and private industry were at the forefront of the development of a new industry to unlock the energy within the atom. Chemical and mechanical processes were adapted from other industries, while methods and structures to handle hazardous radioactive materials were devised from scratch. The industry set new challenges for building contractors to carry out 'civil engineering work to mechanical engineering standards'.[131] To add further pressure, the work was carried out to non-negotiable timetables; for instance the Windscale Piles were completed within a week or so of the target date.[132] Such was the haste that many buildings were started before the final design drawings and plant specifications were complete. In keeping with the age, the architecture of the new atomic factories was utilitarian, distinguished by loose-fitting, steel-framed structures clad in asbestos sheeting and patent glazing. In total the initial nuclear weapons programme required over 1,000 buildings and represented one of the costliest early post-war national construction projects, with the value of the work split approximately between Research Group £14 million, Weapons Group £21.5 million and Production Group £27 million, a total of £62.5 million.[133] All this was achieved against a background of continuing post-war food and material rationing, some of the worst winters in living memory, and constant concerns about the availability of enough skilled personnel. The air-cooled Windscale Piles had fulfilled the immediate demand for plutonium for Britain's first atomic bomb. In the next phase more complex gas-cooled reactors would be developed, also initially for plutonium production, subsequently setting a distinct technological path for the world's first nuclear power programme.

Box 3.2 Protest

Among modern industries, the nuclear industry is exceptional in the sustained hostility it has attracted, paralleled only by opposition to animal-based life sciences. Initially, there was little overt opposition to atomic weapons, or to the location of the places required for their development. The earliest establishments were secretive, generally built within existing or recently vacated defence sites, and as formal planning legislation evolved, most of these remained free from external scrutiny under Crown immunity exemptions. However, even the location and setting of high-priority defence sites at Drigg (Cumbria) and Capenhurst (Cheshire) could be influenced by local amenity concerns (*see* pp. 27, 39).

As the civil nuclear power programme expanded, many of the new stations were built in remote areas noted for their natural beauty, and the intrusion of research establishments, power stations and associated power lines became a common cause of contention. For the new commercial power stations, the amenity clause in the 1957 Electricity Act obliged the Central Electricity Generating Board (CEGB) to take into consideration the impact of new works on their setting, including natural beauty, fauna and flora, and features of architectural and historic interest. At this time, opposition to nuclear sites was generally directed at their visual impact rather than their technologies. At Winfrith (Dorset), there were local worries about the effect of the new research establishment on the Dorset moorlands that inspired Thomas Hardy's 'Egdon Heath', along with fears about water supplies and effluent disposal. At the public inquiry, convened in January 1957, John Cockcroft, director of AERE Harwell, reassured the inquiry that UKAEA would fully consult with local interest groups through the recently established Winfrith Heath Preservation Committee. Characteristically, other views were mixed. Local councils welcomed the employment prospects, as did engineering unions, while employers raised questions about the availability of labour.[1] Where public inquiries were held into the locations of nuclear power stations, local authorities and people were generally optimistic about the economic opportunities they would bring. To allay concerns from local and national amenity groups about the visual impact of the new stations, the CEGB was often required to engage a lead architect and landscape designer, in some cases submitting proposals to the Royal Fine Art Commission.

The close association between the exploitation of nuclear technology with nuclear weaponry has damaged the industry's reputation. This dilemma was acknowledged by the sector's leaders, politicians and their advisors. Viscount Cherwell, physicist, friend and scientific adviser to Churchill, counselled that 'for the first time man has it in his power to destroy all life on this planet, and the question is whether he will be able to avoid some form of universal suicide'.[2] However, for the generations that had fought and lived through two world wars, and against the background of rising international tensions, the prospect of a weapon that might deter future conflicts was an attractive proposition to many. Early opposition to atomic weapons was muted, but from the mid-1950s the development of a new generation of hydrogen bombs and the presence of United States bases in Britain produced more visible opposition.[3] Ambivalence, positive or tacit support remains difficult to recognise. An article in the *New Statesman* in November 1957 by the writer J. B. Priestly may be seen as a point marking the coalescing of individuals and groups opposed to Britain's nuclear weapons. The following spring, around 5,000 people, led by the Direct Action Committee Against Nuclear Weapons, marched in protest over four days the 83km from Trafalgar Square (London) to AWRE Aldermaston (Berkshire).[4] A year later the newly formed Campaign for Nuclear Disarmament (CND) organised the march from Aldermaston to Trafalgar Square, where around 20,000 people attended a rally. The marches lasted in various forms until 1965, with occasional later attempts to revive the tradition.

Other groups advocated taking their protests to remote United States airfields in Britain. In December 1961, Wethersfield (Essex) came to national attention when 600 people from a direct action, anti-nuclear protest group, the Committee of 100, attempted to breach the airfield's perimeter. The 'Wethersfield Six', charged under the Official Secrets Act, received substantial prison sentences for their role in organising the demonstration.[5]

After the mid-1960s, the impetus for large demonstrations waned and the protest organisations shrank in size. A decade later opposition revived with the confirmation of two new advanced gas-cooled power stations, the possibility of a pressurised-water reactor station, and the proposal to build the controversial thermal oxide reprocessing plant (THORP) at Windscale (Cumbria). Previously, public inquiries into the location of new nuclear facilities had focused mainly on amenity issues and were generally short lived. Although these issues remained important, the inquiries into THORP and Sizewell B (Suffolk) were distinguished by complex environmental, economic and technical arguments often led by Friends of the Earth. They also ran for many months: the THORP inquiry lasted for 100 days, while the one for Sizewell B PWR station sat over 340 days and cost £20 million. Nuclear hazards were rarely away from the public consciousness, with the partial meltdown of the Unit 3 reactor at Three Mile Island, Harrisburg (Pennsylvania) in 1979, and more dramatically in 1986 with the explosion of Reactor Number 4 at Chernobyl, Ukraine. It was also a period of increasing Cold War nuclear anxieties following the deployment of Soviet SS20 missiles in eastern Europe in the late 1970s, and the western response of the early 1980s comprising cruise and updated Pershing missiles. The United Kingdom was also preparing to replace its submarine-launched Polaris missiles with the new Trident system, while in the United States, scientists proposed an enhanced radiation weapon, or neutron bomb – deadly to life, but less destructive of infrastructure.

A characteristic of anti-nuclear activity in the early 1980s was the proliferation of ever-shifting local and national groups and protest networks, with individuals aligned to different groups as the focus of opposition changed (Fig. B3.2.1).[6] Inspired by American protest groups and by reactor site occupations by West German and French demonstrators, the new AGR station at Torness (East Lothian) was subject to vociferous opposition from groups such as the Socialist Environment and Resources Association, Friends of the Earth, and the Scottish Campaign to Resist the Atomic Menace (SCRAM). In May 1978, an action organised by SCRAM attracted around 5,000–10,000 protestors who occupied the proposed Torness site. This protest included a march from Dunbar, followed by picnicking and the demonstration of alternative energy sources. Supporters also travelled from England, Holland, Germany and France.[7] The 1986 Chernobyl accident reinvigorated the protest campaign shortly before fuelling was scheduled to begin, and attempts were made to delay the process.[8]

Communities without any previous direct association with the nuclear industry were also drawn into opposition. From 1982 explorations by the government's Nuclear Industry Radioactive Waste Executive (Nirex) into

Fig. B3.2.1
Portskewett, Gwent, about 1980. Mid-Glamorgan CND camp, their roadside posters opposed the proposed local nuclear power station, Sizewell B, nuclear weapons, and US bases.
[Author's collection]

Fig. B3.2.2
Lovers Lane, Thurso, Caithness, 2019. Even in areas with close links to the nuclear industry support was not universal. Protest mural adjacent to the railway station, Caithness Against Nuclear Dumping (CAND), probably late 1980s or 1990s.
[Author's collection]

possible locations for the deep geological disposal of radioactive waste was met with resistance across the country, preventing any field investigations (Fig. B3.2.2).[9] Campaigns against the movement of nuclear waste highlighted the prevalence and possible hazards of radioactive materials, leading some areas to declare themselves 'nuclear free zones' (Fig. B3.2.3).[10]

Acts of protest have left few physical traces, and other than in newspaper archives and the like it is largely an intangible heritage embedded in people's memories. At the sites of temporary protest and peace camps, slight archaeological traces may remain in the topsoil of shelter locations and discarded artefacts, while postcards, metal, plastic and paper badges recall past campaigns (Fig. B3.2.4). Many groups were very active in publishing a rich assortment of pamphlets designed to educate activists. Often produced in small numbers, they reside in private collections and multiple public institutions. A rare permanent monument was built outside a former Royal Ordnance Factory at Elstow (Bedfordshire), which had been identified as a possible location for storing nuclear waste (Fig. B3.2.5). More recently, a circle of Welsh stone and associated monuments have been built at Greenham Common (Berkshire) to commemorate the women's peace camp.

Fig. B3.2.3
Mosley Street, Manchester, 1999. Street sign next to the now removed Manchester Peace Garden.
[© Crown copyright. Historic England Archive, AA99_03274]

Fig. B3.2.4
Postcards with messages opposing the construction of a new power station at Sizewell and nuclear waste disposal at Elstow.
[Author's collection]

Fig. B3.2.5
Elstow, Bedfordshire. The original monument was built in 1986 by Bedfordshire Against Nuclear Dumping (BAND); in 2017, as part of a new road scheme, it was moved from its previous location and restored.
[© Historic England Archive, DP436953]

4

Understanding the atom – government research establishments

Fig. 4.1
Atomic Energy Research Establishment Harwell, Oxfordshire, 8 April 1954. The strong underlying layout of the RAF airfield remains. To the top and bottom are the temporary aluminium housing estates and in the centre the arc of wartime hangars, the Radio-Chemical Laboratory and other new research facilities. [Historic England Archive (RAF Photography), raf_82_895_f21_0251]

Introduction

Since the end of the 19th century, British scientists, and scientists working in Britain, had been at the forefront of the theoretical understanding of the atom, and during the Second World War they were integral to the practical development of an atomic bomb. If Britain was to maintain this leading position, both to develop an atomic bomb and nuclear energy for peaceful purposes, it would require a sustained heavy financial commitment by government to build and support state-of-the-art research facilities. In the early post-war years, the focus remained on commissioning reactor piles and chemical plant able to produce and extract plutonium for use in atomic weapons. From the early 1950s research expanded to include the development of power reactors that might reduce the country's dependency on coal and oil, with the prospect of a lucrative nuclear technology export market. Successive governments remained firmly devoted to the lure of atomic energy, and after the creation of the United Kingdom Atomic Energy Authority (UKAEA) in 1954 its income rose from £47 million in 1954–5 to £106 million in 1958–9, about half of which went on civil applications. By the mid-1960s over a quarter of all government expenditure on civil research and development was spent on atomic power and nuclear physics.[1] This chapter will explore the research establishments built to realise the academic, technological and industrial potential of the atomic age.

Atomic Energy Research Establishment, Harwell

The wartime experiences of British-sponsored scientists working in North America on the Manhattan Project provided first-hand knowledge of what a new atomic research establishment might comprise. In Washington, as early as November 1944, a cosmopolitan group including James Chadwick, John Cockcroft, Herbert Skinner, two Australians Mark Oliphant and Harrie Massey, and a refugee from Nazi Germany, Rudolph Peierls, met to discuss what might be required.[2] The political decision to create a British atomic research establishment was made in April 1945, and in July Sir John Anderson, Chancellor of the Exchequer, wrote to Cockcroft to offer him the proposed directorship.[3] That autumn, on 29 October 1945, the recently elected Labour prime minister, Clement Attlee, announced to the House of Commons that a new research and experimental establishment was to be set up to investigate all forms of atomic energy. Administrative responsibility for atomic research in the Department of Scientific and Industrial Research and the Tube Alloys Directorate was transferred to the Ministry of Supply, Department of Atomic Energy.[4] In November 1945, Cockcroft formally accepted the position as its first director, while retaining his responsibilities in Canada. He was a renowned nuclear scientist, but as importantly had practical firsthand experience of the construction of scientific laboratories at Cambridge and the Canadian Chalk River establishment, while maintaining a personal interest in architecture.[5]

Searches for a suitable site focused on airfields in the south of England that could offer accommodation and basic infrastructure, and after investigating several locations, the late 1930s RAF station at Harwell (Oxfordshire) was chosen (Fig. 4.1). Constructed to permanent airfield standards, it had good-quality, well-spaced brick buildings set among tree-lined roads creating a campus-like atmosphere, with some accommodation available in the officers' mess, barrack blocks and service housing. Its large

hangars also offered ready-made spaces to house scientific plant. In addition, it provided an attractive working and living environment with ready access to academics at the University of Oxford and central government in London. To further enhance the setting, Cockcroft, with his vision for a research campus, invited gardeners from the Royal Botanic Gardens, Kew, to plant rose beds and additional trees.[6]

The Atomic Energy Research Establishment (AERE) assumed control of the airfield on 1 January 1946.[7] Work began almost immediately by the construction firm of Chivers and Sons of Devizes to convert the RAF buildings for new purposes and to create the specialised structures needed by the research establishment. Infrastructure priorities included the first reactors, the Graphite Low-Energy Experimental Pile (GLEEP) and the British Experimental Pile '0' (nought) (BEPO), an electromagnetic separator to produce isotopes and small amounts of U-235, a Van de Graaff accelerator, a synchrocyclotron, and a hot laboratory, as well as general office and laboratory accommodation.[8]

Handling hazardous materials

Before work could begin in earnest, scientific facilities were required that called for a new type of architecture, one designed to minimise the risks involved in handling hazardous radioactive materials. One of the first buildings to be designed by Ministry of Works architects for Ministry of Supply scientists was the Radio-Chemical Laboratory B220 (Fig. 4.2). While the Harwell teams became established, staff in Montreal made an important contribution to its initial planning.[9] It was constructed immediately to the west of the wartime hangars, and when complete was one of the largest structures on site, measuring 115.82m by 76.2m with its tallest point at 32m. One wing of the building was devoted to alpha-emitting materials and the

Fig. 4.2 AERE Harwell, Oxfordshire, Radio-Chemical Laboratory (B220). [Courtesy of the Nuclear Decommissioning Authority]

Fig. 4.3
AERE Harwell, Oxfordshire, Radio-Chemical Laboratory (B220), showing the working face of the caves and the use of manipulators to handle material within the caves.
[Courtesy of the Nuclear Decommissioning Authority]

other to beta/gamma work. To reduce the risk of contamination, the building was divided into individual work areas comprising an entrance vestibule with a remote-controlled wash basin and a hand dryer, or emergency shower, a hand monitor, laboratory, and an office. All radioactive work was done in enclosed containers or glass and stainless steel fume cupboards with a stream of air forming an invisible curtain across the opening. Maintenance of the radioactive areas was carried out from self-contained rear areas with workers protected by adapted wartime rubber 'frog' suits.[10]

In the early 1950s a new wing was added for plutonium work. Greater flexibility was introduced with free-standing glove boxes in a large hall, services to which were carried in the ceiling, including water, vacuum pipes, and electricity, with gas and argon conveyed to the work areas through tubes and cables.[11] By the end of its operational life in 1990, it contained about 120 laboratories and 150 glove boxes, with some 'caves' for handling radioactive materials shielded by concrete walls up to 2m thick (Fig. 4.3).[12] Other safety features in the hot laboratory included smooth corners to prevent the accumulation of dust particles. Its walls were also covered in strip lacquer, which in the event of contamination could be removed and the area resurfaced. The whole building was sealed by double-glazed windows and artificially ventilated, with all intake and outlet air passing through filter banks. Air-conditioning plant which could clean the entire volume of air within the building twice a second was housed in the building's windowless upper storey.[13] When it opened in 1949 it was the world's most advanced nuclear laboratory. Its role as the birthplace of the British nuclear industry was recognised in 2009 when the Royal Society of Chemistry awarded it a Chemical Landmark Plaque. It is currently being decommissioned.[14] The close links with work being carried out at Windscale was illustrated by the first full-scale casting of a plutonium billet on 12 December 1951 in the nearby building B149.[15]

Glove boxes and sealed spaces known as 'caves', which allowed operators to remotely manipulate radioactive materials, are some of the most characteristic features of the nuclear industry. To prevent the escape of their contents, glove boxes were maintained at negative pressure. For the viewing port, or sides of the box for alpha and beta sources, Perspex provided

Fig. 4.4
AERE Harwell, Oxfordshire, Chemical Engineering Building (B351).
[Courtesy of the Nuclear Decommissioning Authority]

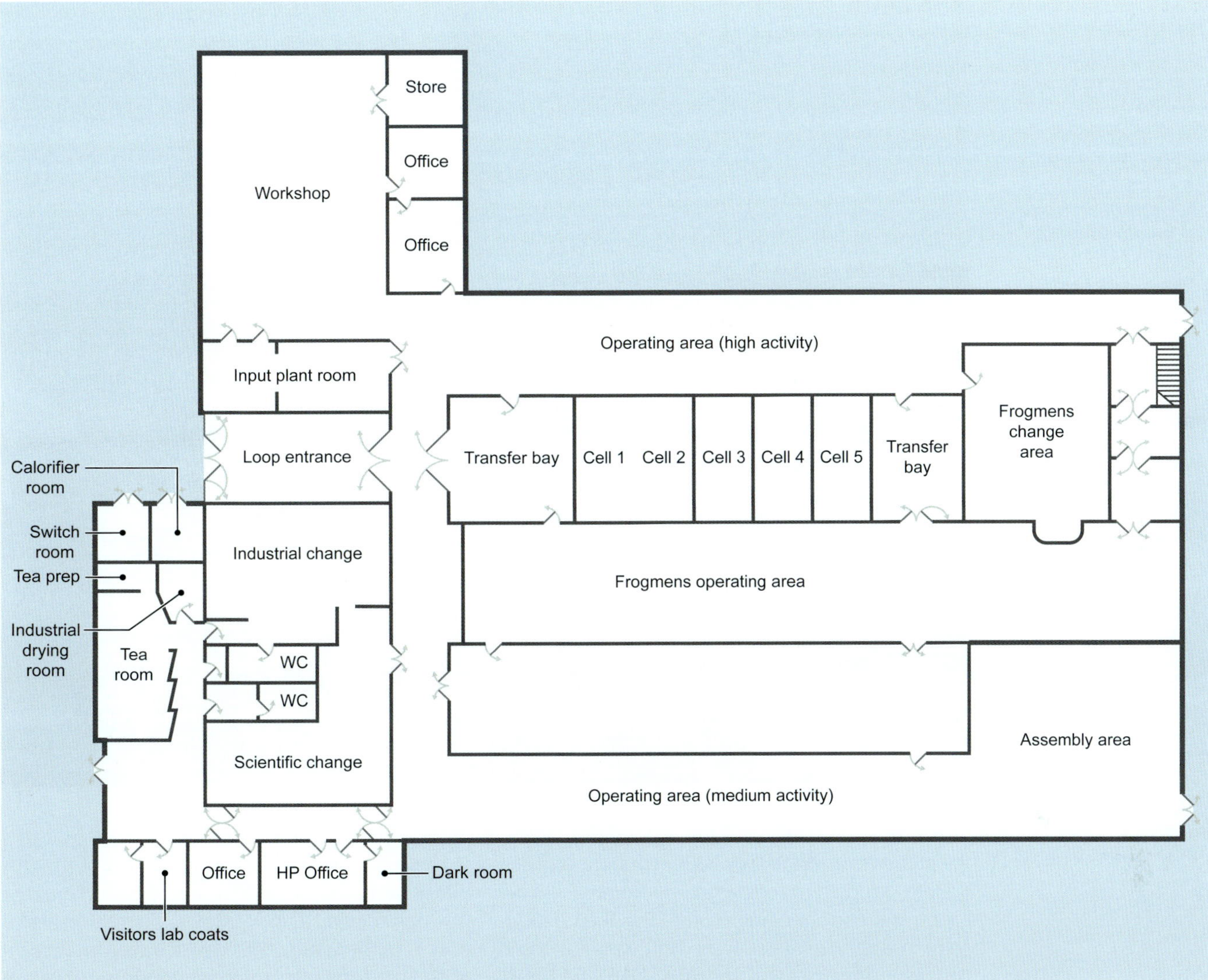

Fig. 4.5
AERE Harwell, Oxfordshire, high-activity handling building (459). [Redrawn from Combustion, Boiler House and Nuclear Review, *February/March 1958, 77)]*

sufficient shielding, while for gamma sources lead-lined glass was used. To handle items operators inserted their hands and arms into long rubber gloves. In other situations, manipulators were used.[16] Initially, chambers, known as 'castles', formed of interlocking bricks, were used, the operators manipulating hazardous materials with tongs viewed through mirrors.[17]

Another early building was the Chemical Engineering Building B351. It stood seven storeys, 27m, tall with a ground plan of 115m by 50m (Fig. 4.4). Steel framed, it was clad in white asbestos sheeting. Internally it provided a flexible open space, 25m high, for the construction of prototype plant or small production runs. The upper storey housed ventilation plant and on other floors were offices and laboratories. The remaining areas were left open and might be used for the construction of temporary lead- or concrete-shielded working areas.[18] Here it was reported Cockcroft's interest in the building's design resulted in the height of its three ventilation stacks being lowered.[19]

In early 1958 a new high-activity handling building, 459, was opened, and this provides a good example of a building engineered to contain highly radioactive materials measuring up to 100,000 curies (Fig. 4.5). The new facility was requested in January 1956 and was designed by a Ministry of Works team including O. P. Carver, senior architect, and senior engineers G. Johnston and J. Greenig; 16 principal contractors were engaged to build and equip it. Its main functions were to disassemble and analyse irradiated fuel and other components from the materials research reactors DIDO and PLUTO (*see* p. 100).

Such was the confidence in the shielded

Fig. 4.6
AERE Harwell, Oxfordshire, high-activity handling building (459). In the foreground the clerestories mark the position of the changing rooms. The shielded handling areas, or caves, were housed in the taller building.
[Courtesy of the Nuclear Decommissioning Authority]

operating cells that along one side of the building stood its ancillary services including offices, a store, workshop, offices, plant rooms, a tearoom, and two changing rooms (Fig. 4.6). These were roughly of equal size but were divided between industrial and scientific staff, the toilet arrangements indicating a solely male workforce. Set perpendicular to this was the main operational area of two heavily shielded blocks of cells, one for high-activity materials and the other for medium activity (Fig. 4.7). The high-activity cells were arranged as a rectangular block of five units, each of which was 2.44m wide, 3.2m deep and 5.03m high, with transfer bays at either end, although in its original configuration the partition wall between cells one and two was omitted. The concrete side walls were 1.68m thick. On the working face, shielding window units, comprising zinc bromide fluid between two panes of glass 1.52m by 0.91m, covered the full width of the wall. For maximum flexibility, the partition walls, 0.61m-thick steel tanks filled with steel-shot concrete, were designed to be moveable. One section weighing 30.48 tonnes could be moved into the maintenance area on electrically driven rollers and the other section moved by hydraulic jacks. Material in the cells was worked on with a pair of manipulator arms controlled by an operator looking through the shielded windows. There was also a power manipulator that could travel the length of the cells through intercell doors to carry out the operations described below. Like the Radio-Chemical Laboratory, maintenance of the cells was undertaken from a sealed rear room only accessible to frogmen in sealed rubber suits, who entered and exited through a dedicated change area (Fig. 4.8). Ventilation was arranged in the building so that the air flow was always towards the most active areas. In operation, material was brought to the loop entrance, equipped with a 25.4-tonne hoist, and moved to the transfer bay. From here it was transferred to cells one and two where it was cut up, in cell three samples were milled and machined for examination, cell four was devoted to metrology, and cell five was used for physical testing. The external finish of the building was functional, with a partly exposed reinforced concrete frame infilled with brick cavity walls. The offices and workshop were lit by windows set in patent steel frames with coloured enamelled steel panels.[20]

Fig. 4.7
AERE Harwell, Oxfordshire, high-activity handling building (459), working face of the caves. [Courtesy of the Nuclear Decommissioning Authority]

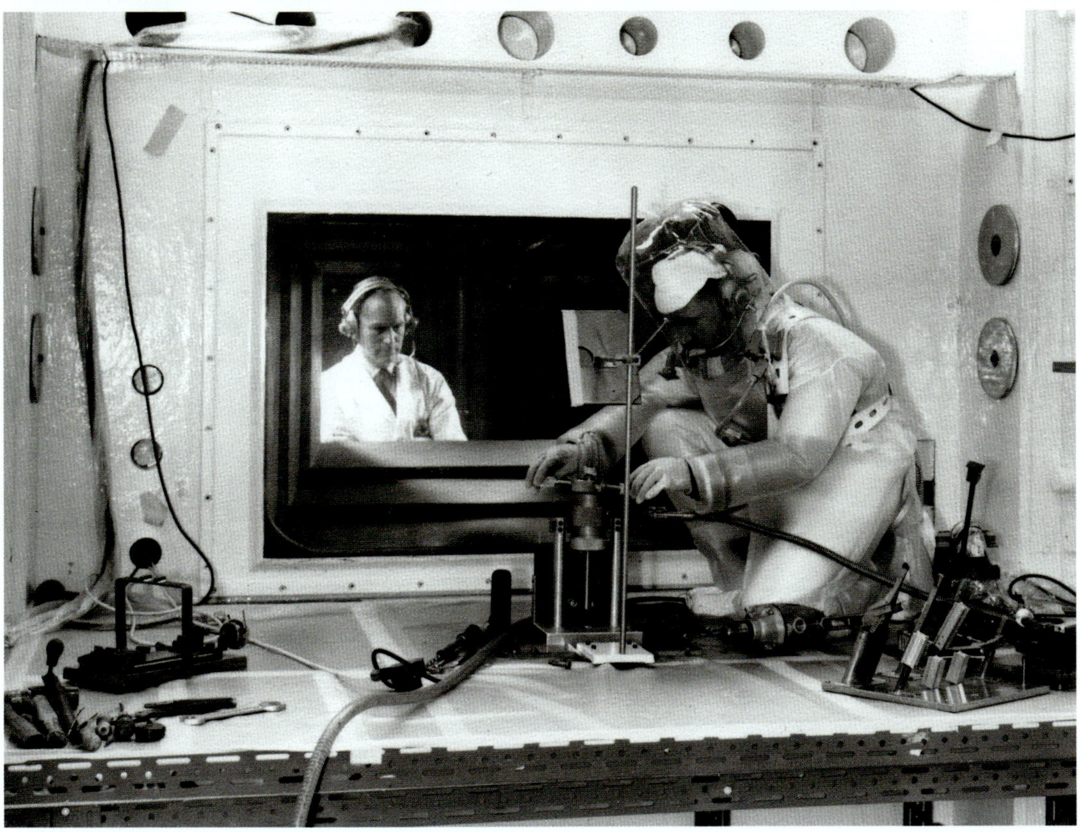

Fig. 4.8
AERE Harwell, Oxfordshire, high-activity handling building (459). Worker in a frog suit in one of the caves. [Courtesy of the Nuclear Decommissioning Authority]

Reactors

In the decade after the establishment's creation, many questions remained about the most efficient and safe combinations of reactor fuels, coolants and moderators, the effects of radiation on reactor structures and other materials, and the handling of waste. To answer these questions, and to gain practical construction and operating experience, 14 experimental reactors were built at Harwell (Table 4.1). Several were also important for isotope production (*see* p. 110). As William Penney later commented, 'Harwell was bursting with ideas about possible reactor systems'.[21]

Soon after AERE acquired Harwell, work began in Hangar 8 on GLEEP, which required a substantial concrete foundation with extensive underground ventilation tunnels (Fig. 4.9). Lessons learnt by the wartime British-Canadian team played a vital role in the design of GLEEP, whose development was led by the New Zealander Professor C. H. Watson-Munro. He had been instrumental in the design of the Canadian heavy-water reactor ZEEP (Zero Energy Experimental Pile), and later went on to become chief of research and development for the Australian Atomic Energy Commission.[22] GLEEP comprised a cylindrical core 5.24m in length and 5.72m in diameter surrounded by an octagonal prism of graphite blocks lying on one of its sides (Fig. 4.10). The blocks weighed 513 tonnes and were fabricated to create a lattice of diamond-shaped channels running through the graphite mass. Due to the shortage of uranium metal, it was fuelled with a combination of natural uranium bars, 30.48cm in length and 2.29cm in diameter, contained in aluminium, altogether weighing 12.19 tonnes, and 21.33 tonnes of uranium dioxide pellets wrapped in paper packages, 30.48cm long and 4.11cm in diameter inserted into aluminium cans. The graphite was surrounded by a barytes concrete shield with walls 1.52m thick and a 1.22m-thick roof, with a 0.91m-wide air gap between the graphite and the wall. The pile was operated by four coarse cadmium control rods and one fine control rod. In an emergency, these rods, acting under gravity, could be used to shut down the pile. Passing through the reactor were seven experimental channels. Initial work in these concerned the properties of materials used in the construction of reactors and isotope production.[23] On 15 August 1947, GLEEP achieved criticality. It was the first successful reactor in western Europe and ran until 1990.[24] In 1960 it was modernised and refuelled with 11,500 uranium cartridges originally intended for use in the Windscale Piles.[25]

Work was also underway in Montreal, and later by the design team at Risley, on the second reactor, BEPO (Fig. 4.11). At Harwell, construction began in June 1946 in Hangar 10 on a deep concrete foundation with substantial subterranean ventilation tunnels; in this case, to accommodate the vertical control rods, the hangar roof was raised by 6.1m. It was a similar design to GLEEP, with a central graphite core surrounded by an air gap with a 1.52m-thick, cuboid biological shield of dense barytes concrete (Fig. 4.12). It was fuelled by 17,600 uranium cartridges weighing 40.64 tonnes that were pushed into horizontal channels and later ejected through the opposite face.[26] The charge face was served by an aircraft carrier-like lifting platform, a feature also found on the later Windscale Piles. BEPO was designed to operate at a much higher power level than GLEEP and required a forced-air ventilation system with filters as the air entered and exited the reactor. Using fans supplied by Metropolitan-Vickers Electrical Company Ltd, the expelled air was vented up a 60.96m circular brick stack, which brought a characteristic symbol of industry to rural Oxfordshire.[27]

For their moderators, both GLEEP and BEPO required substantial quantities of carefully machined pure graphite blocks. During the war considerable experience had been built up in Canada about the manufacture of graphite, and 812.84 tonnes of the 863.64 tonnes of the material used in BEPO was shipped from the Electro-metallurgical Corporation Welland (Ontario).[28] Tests in ZEEP had determined that pure petroleum coke graphite was the best for reactors. Further work was carried out at Harwell, within universities and industry, including Welland, the British Aluminium Company at Kinlochleven in the Scottish Highlands and British Acheson Electrodes, Sheffield. The resulting specification for pure graphite A (PGA) was used in Windscale, Calder Hall and later Magnox reactors.[29] At Harwell, a dedicated graphite machine shop was established which produced around 25,000 blocks in 60 variants for assembly in BEPO. Some of the machinery was later moved to Windscale.[30] The graphite mass was pierced by horizontal and vertical experimental channels, which were mainly used for testing structural and physical

Table 4.1 AERE Harwell reactors and critical facilities.

	Name		*Fuel*	*Moderator*	*Coolant*	*Operational*	*Shut down*	*Maximum heat output*	*Notes*
1	GLEEP	Graphite Low Energy Experimental Pile	Uranium and uranium dioxide	Graphite	Air	1947	1990	3kW	Europe's first successful reactor
2	BEPO	British Experimental Pile O (nought)	Uranium	Graphite	Air	1948	1968	6MW	
3	ZEPHYR	Zero Energy Fast Reactor	Plutonium fuel	None	None	1954		10W	World's first fast reactor
4	DIMPLE	Deuterium Moderated Pile Low Energy	Various	Heavy water	None	1954	1961	10W	Thermal reactor studies, moved to Winfrith in 1960
5	ZEUS	Zero Energy Uranium System	Uranium 235	None	None	1955	1957	100W	Fast reactor studies to study core design for Dounreay
6	ZETR	Zero Energy Test Reactor	Uranium	Heavy or light water	None	1955	1956	Negligible	Moved to Dounreay 1956. Homogeneous aqueous systems
7	LIDO	Thermal swimming pool	Uranium-235	Natural water	Natural water	1956	1974	340kW	Shielding studies
8	DIDO	D2O, chemical symbol for heavy water	Uranium-235	Heavy water	Heavy water	1956	1990	26MW	Materials testing
9	PLUTO		Uranium-235	Heavy water	Heavy water	1957	1990	26MW	Materials testing
10	NEPTUNE	Zero energy	Uranium-235	Light water	None	1957	1959	100W	To study PWR design for submarines
11	NERO	Na (sodium) experimental reactor, 'O' zero energy	Various, slightly enriched uranium	Graphite	None	1957	1959	100W	Moved to Winfrith, renamed JUNO
12	HAZEL (ZETR 2)	Homogeneous assembly zero energy	Enriched uranium	Heavy water	None	1958	1958	1W	Fluoride of uranium trials
13	DAPHNE	DIDO's and PLUTO's Hand Maiden for Nuclear Experiments	Enriched uranium	Heavy water		1962	1967	100W	Converted to DELPHI
14	DELPH I		Enriched uranium	Heavy water		1968	1968	100W	

Principal sources *Electrical Review*, 30 November 1956, 1014; Hance 2006, 71–82. Reactors at the Atomic Energy Research Establishment Harwell, www.osti.gov/servlets/purl/4841165 (acc. 11 September 2023).

materials used in reactors and isotope production. GLEEP operated until 1968.

In July 1954 the DIMPLE (Deuterium Moderated Pile Low Energy) reactor, based on the Canadian ZEEP reactor, achieved criticality. It was a general research reactor offering Harwell scientists direct experience of a heavy-water-moderated system in which different core assemblies could readily be exchanged. The aluminium reactor tank measured 2.59m in diameter and stood 3.05m high. This was surrounded by a graphite neutron reflector which in turn was enclosed by a 0.61m-thick concrete block biological shield with a movable shield over the top of the reactor. The whole facility measured 6.71m square and stood 5.49m tall. In place of control rods, the reactor was controlled by raising and lowering the level of heavy water in the tank. Work carried out in the reactor included the study of heavy-water-moderated core lattices for use in the later DIDO and PLUTO reactors. It was subsequently used with a reactor oscillator for general research. Its final use at Harwell was to carry out initial studies

Fig. 4.9
AERE Harwell, Oxfordshire, GLEEP, the first reactor commissioned in Western Europe on 15 August 1947.
[Courtesy of the Nuclear Decommissioning Authority]

Fig. 4.10
AERE Harwell, Oxfordshire, GLEEP, schematic diagram.

Key
1 Exhaust duct
2 By-pass air extraction system
3 Main air extraction system
4 Stepped seatings for removable concrete blocks
5 Concrete biological shield
6 Winding shafts
7 Coarse control rods
8 Shut-off rods
9 Safety rod
10 Fine control rod
11 Access manhole
12 Catwalks
13 Air inlets
14 Experimental hole (oscillator)
15 Air-flow baffle
16 Control ionisation chambers
17 Access door in air baffle
18 Steel baseplate
19 Graphite uranium lattice
20 Experimental hole L
21 Experimental hole J
22 Experimental hole K
23 Oscillator operating pendulum equipment
24 Ion chambers (gamma monitors)
25 Service duct
26 Control room
[Courtesy of the Nuclear Decommissioning Authority]

Fig. 4.11
AERE Harwell, Oxfordshire, BEPO, in operation in the modified hangar.
[Courtesy of the Nuclear Decommissioning Authority]

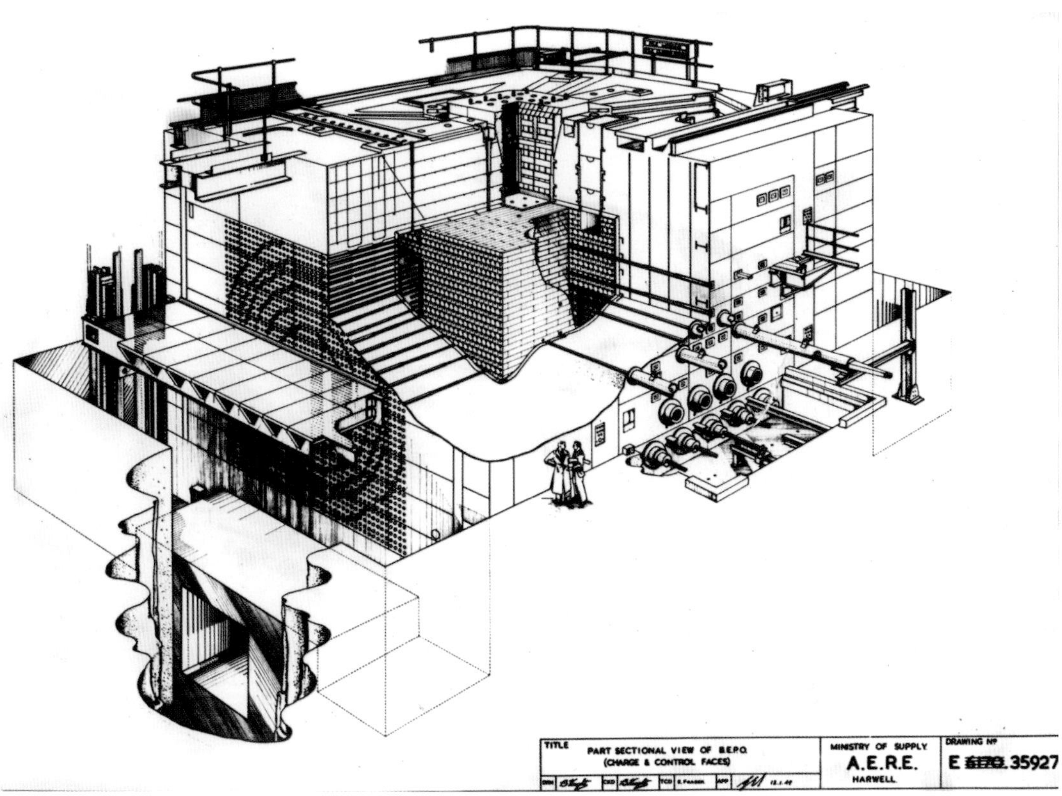

Fig. 4.12
AERE Harwell, Oxfordshire, BEPO, schematic diagram with the loading platform to the left.
[Courtesy of the Nuclear Decommissioning Authority]

into steam-generating heavy-water reactor technology, before it was moved to Winfrith (Dorset) to continue this work (*see* p. 140).[31]

Not all the reactors were intended to directly assist in the design of new power sources, and in the mid-1950s two material testing reactors, DIDO and PLUTO, were added. These were built on the new reactor site, about half a kilometre to the south-west of the main site (Fig. 4.13). They were of a similar design, using fuel of enriched uranium alloyed with aluminium and were moderated and cooled by heavy water; for safety both were housed in cylindrical containment vessels. The facilities were designed by AERE, the Ministry of Works and the North Yorkshire engineering firm Head Wrighton Processes Ltd. To support the work of DIDO and PLUTO, a smaller reactor DAPHNE (DIDO and PLUTO's Hand Maiden for Nuclear Experiments) became operational in February 1962, enabling preliminary experiments to be undertaken without taking up time in the main reactors. DIDO technology enjoyed some export success, and similar reactors were installed at the Australian Atomic Energy Commission research site at Lucas Heights near Sydney, DR3 Riso, Denmark, and JR2 in Germany.[32] A similar reactor was also installed at Dounreay (Caithness) (*see* p. 115).

In 1957 NERO (Na (sodium) Experimental Reactor 'O'), a zero-energy reactor, came into operation, so called because it operated at such low power that no coolant was required. This was a relatively small reactor, about 4.27m in diameter and 3.05m high, surrounded by a 0.61m-thick concrete shield. Its main purpose was to investigate issues with more advanced types of graphite-moderated reactors, including cooling with sodium.[33] It was also later moved to Winfrith (Dorset).

Fast reactors

The uranium-fuelled, graphite-moderated, and air-cooled thermal piles discussed so far had met the immediate goals of providing the know-how to construct and operate the plutonium production piles at Windscale and the first generation of power stations. However, in the late 1940s and early 1950s there were growing concerns about the security of uranium supplies, and with the recent world war in mind, apprehension about the vulnerability of sea routes.

Fast reactors use plutonium as a fuel, which is produced as a by-product of burning uranium fuel in thermal reactors. There was the expectation that after the weapons programme had fulfilled its needs, plutonium would become readily available for civil applications. By using

Fig. 4.13
AERE Harwell, Oxfordshire. To the bottom right is the reactor site with the containment vessels for DIDO and PLUTO; to the rear is the massive diamond light synchrotron building; and to the left in the far distance the now demolished Didcot power station.
[© Historic England Archive, 26488/002]

plutonium as a fuel without a moderator, the chain reaction may be sustained mainly by fast neutrons. The absence of a moderator has the benefit of permitting a more compact core with a higher heat rating, that is assuming a suitable coolant can be devised. In this design, excess neutrons are absorbed by the U-238 in the core and a surrounding blanket of uranium, which in turn 'breeds' more plutonium.[34]

One possible research avenue was to use fast reactors to convert non-fissionable U-238 to the fissile isotope of plutonium-239. The Ministry of Supply's Power Steering Committee recommended a dual-track approach of pursuing research into well-understood thermal reactors as well as theoretical fast-breeder technology. The first Fast Reactor Design Committee met on 9 October 1951, when Christopher Hinton argued for a real fast-breeder project rather than an extensive trials series.[35]

Typically, in the early programmes, design and research often ran in parallel, with the construction of new facilities, and in the case of the Dounreay fast reactor (*see* p. 116), construction began at about the same time that the first experimental fast reactor was being built at Harwell. To support the fast-reactor programme, J. V. Dunworth had started work in 1952 designing ZEPHYR (zero energy fast reactor), which went critical in early 1954. It had a 15cm-diameter core of clad plutonium rods surrounded by a natural uranium rod blanket; control was achieved by withdrawing some of the uranium rods. It was Europe's first plutonium reactor and demonstrated that for one atom of fissile material burned, two could be created.[36]

This was followed in 1955 by ZEUS (zero energy uranium system) which was built to gain an understanding of the operation of a fast reactor and to check the calculations on which the fast reactor being built at Dounreay were based. Its cylindrical core was about 50.8cm in diameter and an identical measurement in height and was formed of enriched uranium, provided by Capenhurst. This was surrounded by many tons of uranium in which plutonium formed during its operation. The reactor was operated at low power, obviating the need for cooling. Associated with this programme was the ZETR (zero energy thermal reactor), which was designed to study the use of nuclear fuel in solution using plutonium, U-235 and U-233.[37] It spent little time at Harwell and was moved to Dounreay in 1956.[38]

Thermonuclear, or fusion, power

The reactors discussed so far relied on the principle of fission, the release of energy caused by the splitting of heavy atoms, such as uranium or plutonium. An alternative was to harness the energy produced by joining, or fusing, the atoms of light elements, such as hydrogen: in essence, the source of energy for the sun and the stars. The challenge was to discover methods to heat gases to temperatures equivalent to that of the sun and to devise ways to contain this superheated plasma stream. Theoretical research into this had begun before the war, with work undertaken at Imperial College London and the universities of Liverpool and Oxford. Imperial College moved their work to AEI Aldermaston (*see* p. 113, Box 4.2), while between 1950 and 1952 research at the Clarendon Laboratory, Oxford, was transferred to Harwell, where it was the responsibility of the General Physics Division, led by D. W. Fry. The focus of this work was Hangar 7, where the Harwell Mark III torus was installed. This comprised a ring-shaped field coil with a 34cm core. Based on this machine, in 1955 design studies began on a far larger experimental machine ZETA (zero energy thermonuclear assembly) (Fig. 4.14). This was also installed in Hangar 7 and started up on 12 August 1957.[39] At the time it was the most complex fusion research machine in the west. Fusion research also began at AWRE Aldermaston in Building N60, where a device named OSWALD, comprising a short discharge tube and a bank of 144 condensers, was able to produce a current of 240,000 amps. Work in this building also attempted to raise deuterium to its burning point of about 1 million degrees Celsius.[40]

The main elements of the ZETA assembly comprised an aluminium tube, or torus, with a bore of 1m and a mean diameter of 3m, formed into a doughnut shape containing deuterium gas. This was connected to an industrial high-frequency generator that was used to initiate the electrical discharge in the gas. An innovation at Harwell was to wind the torus with thick electrical cabling through which a magnetic field could be induced to stabilise the ionised gas plasma stream inside.[41] The torus passed through two large ring core pulse transformers made from special-grain oriented steel, weighing 152.4 tonnes, surrounded by coils of electrical cables through which a pulse of electricity was released to create a discharge in the gas inside the torus and so heating it. Two glass sections in the torus wall permitted observations. Associated electrical plant included a transformer, charging generator, condenser bank and switch. The heavy electrical plant was supplied by Metropolitan-Vickers Electrical Company with support from AERE's Engineering Division.

The challenge of achieving controlled nuclear fusion was to overcome the strong electrostatic force of positively charged electrons repelling the nuclei as they came together. The solution proposed was to make the atoms move so fast that the energy of motion would overcome the forces of resistance. To achieve this would require phenomenal temperatures measured in millions of degrees Celsius. After the air had been pumped out of the torus it was partly filled with rarified deuterium gas; an electrical current was then applied to the gas which set up a discharge within it making the gas very hot. If sufficient current was then applied, some of the atoms would fuse, with a corresponding release of energy and neutrons. The current also caused the gas to act as if it was surrounded by a strong magnetic field, which had a pinch effect, constricting the gas into a narrow column and causing its temperature to rise further, a type of containment that has been described as a 'magnetic bottle'.[42] This effectively separated the gas from the wall of the torus, while heavy electrical cabling wound around its exterior created a magnetic field to guide the stream of hot gas, or plasma, within the torus aided by eddy currents in metal walls.[43]

The understanding of fusion had a possible crossover into weapons technology and so the work at Harwell, and parallel research in the United States, was kept secret until it was revealed in a special issue of the scientific journal *Nature* in January 1958. The article noted that neutrons had been observed, one marker that fusion had been achieved. At an accompanying press conference, when pushed by reporters, Sir John Cockcroft unwisely stated that he was 90 per cent certain they were the result of fusion. This was further simplified by newspaper headlines that fusion had been achieved. Ignominiously, a few months later a statement was released clarifying the earlier claim.[44]

Fig. 4.14 AERE Harwell, Oxfordshire, ZETA, showing the horizontal wire-wrapped torus passing through one of the ring-shaped core pulse transformers. [Courtesy of the Nuclear Decommissioning Authority]

ZETA had contributed to the understanding of fusion technology, but far higher gas temperatures would be needed for a sustained period to achieve the goal of power generation. In 1958 fusion research was brought together in the Controlled Thermonuclear Reactions Division, with initial research being undertaken at Harwell before being transferred to Culham (Oxfordshire) in the early 1960s (*see* p. 146). ZETA ran until September 1968, after which it was dismantled; a model and a section of the machine are preserved by the Science Museum.

Big science

Cutting-edge scientific endeavour was no longer characterised by small pieces of elaborately handcrafted wood and brass laboratory equipment, as it had been before the 1930s, but now relied on the expertise of commercial heavy electrical and engineering firms. In addition to the reactors, in the late 1940s work also began on the means to accelerate charged particles, such as electrons or protons, to very high speeds. At Harwell, in H8, a Cockcroft-Walton particle accelerator was installed along with a Van de Graaff generator; another was housed in a former RAF navigation training building. In Hangar 7 a huge hole, in part to act as shielding, was excavated and filled with concrete to support a 3m-diameter cyclotron, its magnet alone weighing 711 tonnes.[45] A much larger tandem Van de Graaff generator was designed with two machines mounted in series and installed in the late 1950s in building B477 (Fig. 4.15). This comprised a range of single-storey offices and a semicircular range surrounded by a brick wall for workshops and experimental areas (Fig. 4.16). At its centre was a reinforced concrete tower with 1m-thick walls containing a pressure vessel 14m high and 3m in diameter which housed the generator (Figs 4.17 and 4.18). This could produce millions of volts as well as itself serving as a particle accelerator. On the side of this was a steel-framed, aluminium-sheet-clad service tower with stairs and a crane. The equipment was supplied by the Metropolitan-Vickers Electrical Company Ltd and ran until 2002.[46] Similar generators were also installed at Aldermaston and the Daresbury Laboratory (Cheshire). Around the same time, Metropolitan-Vickers also supplied a travelling wave linear accelerator for the Neutron Project, spread over 8ha to the south of the Harwell airfield.[47] The English Electric Company, Stafford, was also quick to seize the potential of the new technology, and in the 1940s established a nuclear physics section in a former aircraft engine test plant at Blackheath, 5km from Stafford. Early products included Van de Graaff generators for Harwell and the Cavendish Laboratory.[48]

Fig. 4.15
AERE Harwell, Oxfordshire, showing the tall Van de Graaff generator building and to its rear the RAF hangars. The one to the left with the raised roof housed the reactor BEPO, and to its left is the Radio-Chemical Laboratory. To the right is the airfield control tower converted into a medical centre.
[Courtesy of the Nuclear Decommissioning Authority, HLC 73699]

Rutherford High Energy Laboratory

In 1957 the National Institute for Research in Nuclear Science (NIRNS) was established by royal charter. Its principal role was to provide research facilities for government establishments, universities and others, where the acquisition of hardware was beyond the means of a single institution. Its mission also included the training of scientists and engineers, research into specific topics and the dissemination of knowledge.[49] Accommodation for the new institute was built to the south of AERE's Harwell complex, beyond the secure perimeter fence, reflecting its open academic philosophy (Fig. 4.19). The level airfield presented a blank, flat canvas, and the line of one of the runways provided the orientation for the new establishment. Initial architectural drawings were produced by G. W. Dixon, chief architect of the UKAEA's southern works organisation.[50] The buildings were typical of other Authority locations, with no office or laboratory blocks more than three storeys in height. They presented a clean, modern, purposeful appearance with horizontal glazing bands providing abundant natural light and infill panels coloured in pastel blue or buff with doors picked out in lemon yellow. At the outset its principal equipment included a 50MeV proton linear accelerator, which was transferred by the UKAEA in 1959, and a 7GeV proton synchrotron, known as NIMROD. The institute also established the Atlas Computer Laboratory (ACL) to house a room-sized Ferranti Atlas 1 computer, manufactured at West Gorton,

Fig. 4.16
AERE Harwell, Oxfordshire. Van de Graaff generator building under construction, about 1960.
[Courtesy of the Nuclear Decommissioning Authority, A6408]

Fig. 4.17 AERE Harwell, Oxfordshire. Model of the Van de Graaff generator showing the pressure vessel. [Courtesy of the Nuclear Decommissioning Authority, 050042/27]

Fig. 4.18
AERE Harwell, Oxfordshire. Van de Graaff generator experimental area.
[Courtesy of the Nuclear Decommissioning Authority, HP8754]

Manchester. From 1964, funded through the Science Research Council, ACL became a national computing service and ran until 1973.[51]

Other facilities

In common with many other post-war government research establishments, Harwell was supported by a wide variety of in-house services including engineering, electronics, design offices, photography, computing, and in this case specialist maintenance personnel for the radioactive areas. Some activities were kept within the establishment to maintain confidentiality, while other tasks were highly specialised and unavailable commercially. In some fields, such as the development of radiation monitors for civil defence, the services and industry, initial design work was undertaken at Harwell, with production then contracted out to the private sector, often initially to the radio industry.[52] As knowledge of nuclear technology developed, new industries grew which were capable of designing and producing radiation monitoring equipment and protective clothing. In 1948, to support the development of specialised electronics, a team was transferred to Harwell from the wartime Telecommunications Research Establishment, Malvern (Worcestershire).

Calculations and data analysis were central to Harwell's activities. At first, they were undertaken with logarithm tables, slide rules and desk-top hand-cranked and electric calculators, the latter operated by mainly female teams. Through the 1950s these activities were further automated, beginning with punch-card machines supplied by the British Tabulating Machine Company based in Letchworth (Hertfordshire). In the early 1950s a notable achievement by the Electronics Division was the construction of the world's first all-transistor

computer CADET, which operated between 1951 and 1958. As Britain's computer industry grew, Harwell was able to acquire a room-sized Ferranti Mercury computer in 1958. This was followed in 1961 by a Ferranti Atlas computer which needed a purpose-built building, R31. To install this machine, 19 lorry-loads of components were required, and when operational it was reputedly the most powerful computer in the world; it operated until 1973. From the early 1960s the ascendancy of the American computer industry was evident, with the purchase of IBM mainframes, and during the 1980s Cray supercomputers were purchased, culminating in 1987 with a Cray2.[53] But, the mainframe era came to an end within a decade, and personal desktop computers were to become the norm.

UKAEA Harwell also operated a number of outstations, including the Radiochemical Centre at Amersham and the Radiation Laboratory at Wantage (Oxfordshire) (*see* Box 4.1). In 1956, extra engineering and electronic manufacturing capacity was provided by a factory in the new town of Bracknell (Berkshire). This closed a decade later, a casualty of the mid-1960s contraction in UKAEA's research programmes.[54] More controversially, it also operated an outstation at the Royal Arsenal, Woolwich, involved in human tissue research, including detecting strontium-90 in bone; this laboratory closed in 1966 when new facilities were opened at Harwell.[55]

Harwell's legacies

The period from the establishment's opening in 1946 until the early 1960s has been seen as Harwell's golden years. A relatively young workforce had been brought together to explore new sciences, in the process also building new communities. They had investigated the fundamental scientific and technological questions that enabled the creation of the British nuclear industry. By the early 1960s the civil nuclear power programme was slowing down and much fundamental knowledge about reactor types, operations, and relative merits of different options were understood. After the feverish reactor building projects of the late 1940s and 1950s, facilities built from the late 1960s reflected a change in emphasis to more basic science research. In government there was also a desire to see state research establishments contributing more to modernising British industry, leading to the creation by the new Labour government of the Ministry of Technology in 1964. With greater flexibility in its work programmes, UKAEA began research into topics as diverse as the desalinisation of sea water and the hydrostatic extrusion of metals.

Fig. 4.19
AERE Harwell, Oxfordshire, 1967. In the middle of the airfield the new Rutherford Laboratory is clearly visible centre left, bounded by two former runways. To the right is one of the post-war estates of aluminium bungalows and below, the rectangular wartime bomb stores. [Historic England Archive (RAF Photography), raf_58_8107_f21_0399]

HYDROGEN
NO
SMOKING

Box 4.1 Amersham Radiochemical Centre

In the late 1940s, the commissioning of reactors at Harwell enabled the development of an important new industry to advance the use of diagnostic radioactive isotopes in medicine, industry and other research. Initially this work was carried out by the Isotope Division at Harwell using targets irradiated in the BEPO reactor, and later, if a longer irradiation time was required, at the Windscale Piles (Cumbria).[1] After the closure of BEPO in 1968, other Harwell reactors DIDO and PLUTO supplied isotopes until their shutdown in 1990.

In 1950, as isotope work expanded, it was decided to move the focus of this activity to Amersham (Buckinghamshire). The site chosen was to the east of the town in Little Chalfont (Fig. B4.1.1). In early 1940, Chilcote House, a detached early 20th-century villa and its grounds, had been bought by Thorium Ltd for processing naturally radioactive radium for the manufacture of luminous paint for gunsights and instrument faces. It also produced radioactive material for academic laboratories and neutron sources for the determination of neutron yield in uranium fission. In September 1946, the Ministry of Supply purchased the works for £40,000, but it continued to be managed by Thorium Ltd as an agency factory. It became the national centre for the manufacture and distribution of radium, radon and artificial radioactive substances for scientific, medical and industrial purposes. With increased activity by the end of the decade, its staff rose to 70, more land was acquired, and new laboratories were constructed.[2]

Fig. B4.1.1
The Radiochemical Centre, Amersham, Buckinghamshire. Pictured in 1968 from the north-west. To the right, fronting onto White Lion Road, is a post-war office and to its rear gabled huts dating to the 1940s, with Chilcote House set in the trees to the left. To the rear of the site are the laboratories added during the 1950s and 1960s.
[© Historic England Archive (Aerofilms Collection), EGF1443]

In May 1949, it changed its name to the Radiochemical Centre, and in the following year became part of the Atomic Energy Research Establishment (AERE), effectively an outstation of Harwell. The latter's association with isotope work continued through the Harwell Isotope School, established in 1951 to train scientists and engineers in the application of this new technology, and the Isotope Production Unit, built in 1959. In that year, training activities were transferred to the newly established Wantage Radiation Laboratory (Oxfordshire), which also included production facilities including a package irradiation plant and the gamma sterilisation of medical instruments (Figs B4.1.2 and B4.1.3).[3] This was built to the north of the town on the western side of Grove airfield and was relatively short-lived, closing in 1970.

With the move to the AERE, the emphasis of the work at Amersham shifted from natural radium compounds to isotopes produced in reactors. Over the next few years, the manufacturing facilities were considerably expanded and in 1954 the country's first buildings dedicated to the production of isotopes for industrial and medical uses were opened. Important work was carried out in developing organic compounds using carbon-14, one of the most widely used research isotopes. The 1950s saw increasing demand for use of isotopes to monitor production processes and a corresponding expansion of facilities at Amersham. In 1957, a building with enclosures to enable personnel to remotely manipulate radioactive substances was inaugurated, followed by further laboratory capacity to handle carbon-14 and tritium chemical labelling work. By 1960, the workforce stood at around 450. In the following decade, a further 1,394 sq m of space for organic and inorganic laboratories was added. This included the glass-fronted offices and laboratories by J. Douglass and Partners that were added in 1964, with a large atomic model physics sculpture in its courtyard.[4] Another important technological advance during the mid-1960s was the world's first installation of a cyclotron for isotope production,

Fig. B4.1.2
Wantage Radiation Laboratory, Oxfordshire. The Isotope School in September 1964. [Courtesy of the Nuclear Decommissioning Authority, OP362374]

which operated by bombarding metallic targets with high-velocity electrically charged atomic particles.[5] Amersham became an international leader in the supply of isotopes and a major exporter, with personnel numbers rising in the mid-1970s to around 1,000 (Fig. B4.1.4).

By this time AERE's activities had outgrown the Amersham site, and from 1976, construction work started on a 12ha greenfield site in Cardiff, although work was delayed pending a

Fig. B4.1.3
Wantage Radiation Laboratory, Oxfordshire. Building 500, Laboratory C.49 interior. [© Crown copyright, P_G07175_001]

Fig. B4.1.4
The Radiochemical Centre, Amersham, Buckinghamshire. Radiation Sources Department production centre used for the rolling of alpha foil. Note the use of glove boxes and clean-area working clothing. March 1980.
[Source: The Radiochemical Centre, AJ8185c]

planning enquiry.[6] The new facility, known as The Maynard Centre, housed the Custom Preparation Group, opened in 1980. Employing about 400 people, and responsible for about half of the company's production, the centre specialised in the life sciences and the supply of carbon-14 and tritium products.[7] It closed in 2010. In 1976, laboratories were also established in Gloucestershire for freeze-drying non-radioactive diagnostic products.[8]

The company came to national attention when, on 17 February 1982, it became the first public undertaking to be privatised by the Conservative government led by Margaret Thatcher, and took the name Amersham International plc. A new group headquarters was opened at Amersham Place, Little Chalfont, and in 1985 facilities for non-radioactive technology were built at nearby Pollards Wood. The late 1990s saw several changes to the firm's ownership, and in 2004 it was sold to the American company GE Healthcare.[9]

In February 2005 the Royal Society of Chemistry awarded the Grove Centre, named after Amersham's founder, the wartime chemist Dr Patrick Grove, the status of Historical Chemical Landmark. It was recognised for its pioneering work in the life sciences and as a major centre for the development of the medical and industrial applications of radioactivity in the United Kingdom during the past 60 years.[10] Large areas of the site are currently being cleared for redevelopment.

Isotopes definition

Isotopes of an element have the same number of protons in the atomic nucleus, and therefore have the same chemical properties. They have different numbers of neutrons and therefore have different atomic weights and physical properties. For example, U-235 and U-238 are isotopes of uranium.[11]

Box 4.2 Industrial and educational research reactors

In addition to the large government research centres, several industrial companies and universities also maintained small research reactors. These were used to develop new manufacturing and diagnostic techniques for atomic research, and for training students and technicians to work with nuclear reactors.

In 1947, Associated Electrical Industries (AEI) established a research establishment at Aldermaston Court (West Berkshire), just to the north of the wartime airfield and later Atomic Weapons Research Establishment (AWRE). AEI's activities included research into metallurgy, semiconductors, the physics of surface and wear, and nuclear sciences, and from an early date it was equipped with two Van de Graaff accelerators. AEI also had early involvement in fusion research when George Thomson of Imperial College London moved his research to Aldermaston in 1951. By the end of the decade, they were working in conjunction with UKAEA Harwell using a torus *Sceptre III*

Fig. B4.2.1 Associated Electrical Industries Research Establishment, Aldermaston Court, Berkshire.

Key
1 Reactor building and control room
2 Water demineralisation plant
3 Subcritical facility and fuel-element store
4 Effluent treatment plant
5 Pump house
6 Offices and laboratories
7 Aldermaston Court
8 Oxford House

[Contains OS mapping. © Crown copyright (and database rights) 2025 OS AC0000815036]

incorporating a 30.48cm tube with a mean diameter of 1.14m threaded through an iron core weighing 4.06 tonnes; this was later superseded by *Sceptre IV* (*see* ZETA, p. 101).[1]

In 1956, in association with John Thompson Nuclear Energy Group, AEI began the construction of Britain's first private research reactor. The facility comprised a main building housing the reactor, experimental area, control room, the fission product detector and ventilation plant. Closer to the lake, which was used for secondary cooling, were the primary and secondary coolant pump houses, a water demineralisation plant and effluent treatment facilities. Other buildings included an active storage area, boiler house, a subcritical laboratory and other offices and laboratories (Fig. B4.2.1).[2] The reactor, MERLIN (medium-energy research light-water-moderated industrial nuclear reactor), was of the 'swimming pool' type, where the moderator, coolant and reflector were water. It was fuelled by uranium oxide-enriched uranium-235 and clad in aluminium.[3] It was housed in a vertical aluminium tank, the upper section of which was 3.5m in diameter and the lower section 1.65m in diameter. The whole was shielded by concrete – the lower section by barytes concrete and the upper section by Portland cement concrete. An unusual feature of this reactor was the facility to move the core vertically. The top position was used for handling fuel and experimental elements, the two middle positions were opposite to experimental ports, and the lower position was used for the safe storage of the reactor.[4] Fuel loading began on 6 July 1959 and the reactor went critical later in the month. It operated until about 1965, and subsequently the facility has been demolished.

By 1959, the Hawker Siddeley Nuclear Power Company, near Slough (Berkshire), had designed and built a small research reactor, *Jason*. This was a type of American training reactor in the Argonaut series, ARGOnne Nuclear Assembly for University Training. It was relatively small and could fit into a room 14.6m by 11m. It was envisaged that it would be used to produce medical and industrial isotopes, as well as being available for training engineering and chemistry students. It was water-moderated, cooled with graphite reflectors and fuelled by uranium-235 sandwiched in aluminium plates.[5] Hawker Siddeley operated it for a couple of years until it was acquired by the Department of Nuclear Science and Technology in the King William Building at the Royal Naval College, Greenwich, for training naval submarine personnel. There, it went critical in 1962. It was the first training reactor to be installed in a United Kingdom educational establishment, and it remained in use until the closure of the college in 1998. Its control panel, by Elliott-Automation Limited, Borehamwood (Hertfordshire), is preserved by the Science Museum. Further training for naval students was also available at the Admiralty Reactor Test Establishment, Dounreay, where a prototype pressurised water reactor was installed.[6]

Small research reactors were also installed by leading industrial companies. At Derby, Rolls-Royce operated a pressurised water reactor associated with their nuclear submarine propulsion work. Imperial Chemical Industries (ICI) installed an American General Atomics TRIGA pool-type reactor at their Billingham works, Stockton-on-Tees, that ran between 1971 and 1996.[7]

In 1961, the government announced a programme to provide a small number of research reactors for use in British universities. The first group of reactors was supplied by Advanced Technology Laboratories (California). These were water moderated with graphite reflectors, and used enriched uranium fuel. Examples of this type were supplied to the Universities Research Reactor, Risley (Cheshire); the Scottish Universities Research Reactor Centre (East Kilbride); and Queen Mary College (London).[8] At the last it was installed for a short time between 1964 and 1966 in Mile End Road. It was then moved to the new Nucleonics Laboratory in Marshgate Lane, where it operated from June 1968 until April 1982. In 1970, it was joined by the HELEN facility, which was used for the study of enriched uranium fuel; this ceased operation in 1973.[9] Imperial College London constructed a small reactor centre with associated radiochemistry and physics laboratories, workshops, office and study space on its Silwood House campus at Ascot (Berkshire). Its reactor, CONSORT II, which went critical in April 1965, was British built and designed by the Nuclear Power Group, Imperial College, and GEC (Engineering), Erith (Greater London). It operated until 2008. It was a swimming pool type housed in a vertical cylinder, where again, water acted as the moderator, reflector and coolant.[10]

Dounreay Experimental Research Establishment

In January 1953, the intention to move from the experimental fast reactor ZEPHYR at Harwell to a larger prototype was announced to the House of Commons by the Minister of Supply, Duncan Sandys.[56] Initially, it was proposed to build the experimental fast-breeder reactor at Windscale, but concerns about the safety of an untried design led to the search for a more remote site.[57] Criteria for a location included a coastal site to provide cooling water, but with sufficient elevation to avoid flooding, suitable bedrock, and a low surrounding population density. It should also be available for immediate occupation. The site selected at Dounreay (Caithness) was a wartime RAF Coastal Command airfield, latterly occupied by the Royal Navy as HMS *Tern III*.[58] The airfield and adjacent farmland were formally acquired by the newly formed UKAEA in 1954. Construction work began early in the following year and reached completion on 8 December 1958 (Fig. 4.20). Off site, 11km to the south, a dam was built across Loch Shurrery to provide raw and treated water for general site uses.

Over 600 contracts were let for the construction and fitting out of the new establishment at a total cost of £28.5 million, providing employment for over 2,000 construction workers. The main civil engineering contractor was Whatlings Ltd of Glasgow, and the Motherwell Bridge & Engineering Co. Ltd was responsible for the fast-breeder reactor sphere and steel containment vessel of the materials testing reactor. Alex Findlay & Co. Ltd, also of Motherwell, supplied the structural steelwork as it had done at Windscale. Such was the demand for bricks, the project established its own brickworks to the south of the airfield adjacent to the airfield's accommodation site. These were simple sand and cement bricks, and their production was the one manual construction task that employed a largely female workforce. Precast concrete components were supplied from a yard in Thurso established by George Bruce & Co. Ltd.[59] The airfield camp buildings, sections of which were named after wartime aircraft, and an old army camp at Thurso, provided accommodation for

Fig. 4.20
UKAEA Dounreay, Caithness, August 2013. Its high-tech facilities set in a rural landscape above the rocky foreshore present a sublime image of science and wild nature. In the foreground are the low-level waste vaults, and beyond, the white containment vessel of the materials testing reactor and the dome of the fast-breeder reactor. At the far end of the site are the Prototype Fast Reactor and the Vulcan Naval Research Test Establishment. [Courtesy of the Nuclear Decommissioning Authority, IMG 9805]

construction workers, and the control tower became the main site office (*see* Fig. B6.1.2).[60] The establishment was designed by the Industrial Group of UKAEA at Risley under the chief architect R. S. Brocklesby, with A. V. Levens responsible for work on site. It was claimed that buildings were kept low to minimise their visual impact, also necessary to avoid the battering from high winds.[61]

As with other recently abandoned defence sites acquired by UKAEA, the airfield's layout influenced the design of the establishment. The entrance to the site was along the eastern subsidiary runway and the main buildings were laid out to a rectilinear plan that followed the south-west to north-east orientation of the main runway, set between it and the clifftop. The RAF requested that sufficient distance was left adjacent to the runway for it to remain viable in the event of war.[62] As you enter Dounreay along the former eastern subsidiary runway, the sphere of the fast reactor dominates the skyline, positioned at the notional intersection of this runway and the northern end of the western taxiway. Until early this century the concrete paving of its adjacent airfield dispersal points remained visible; today two hangars survive. The easterly end of the main runway remained in use for UKAEA staff flights until the 1990s.[63] After completion, the permanent establishment was around 2,500 people, 400 of whom were professional scientists and engineers.[64]

Dounreay Fast Reactor (DFR)

The most distinctive feature of Dounreay, and an industrial form that has come to symbolise the industry, is the sphere of the fast-breeder reactor (Fig. 4.21). Despite its familiarity, spherical containment vessels in the nuclear industry are unusual internationally, and in the United Kingdom the only other examples are the Windscale advanced gas-cooled reactor and the dome of the Sizewell pressurised-water reactor. The fast-reactor group at Dounreay comprised the reactor, housed in the sphere, and to the west the heat exchanger building and turbine hall with adjacent diesel generators to provide electrical power when the reactor was offload (Fig. 4.22). On the foreshore to the north were the pumphouse to supply cooling water and the outfall for used water. The two-storey administrative building to the east also housed the octagonal control room; its control desks and associated instrumentation have been saved by National Museums Scotland. To its rear was the active element storage building dominated by an 18.23m vent stack, and the active element storage pond. In common with many other UKAEA sites, these ancillary buildings are simple steel portal frame structures, with steel-framed windows set in horizontal bands, while the remainder of the building was covered by sheet cladding. The only concession to the local vernacular was the use of Caithness stone for some gable walls and the cement bricks made from local sand.

Fig. 4.21
UKAEA Dounreay, Caithness, fast-breeder reactor, 1962. To the right are the control and administration, and the fuel storage buildings. In the centre is the dome enclosing the reactor and to its left the multistorey heat exchanger and stand-by diesel generator buildings. [Courtesy of the Nuclear Decommissioning Authority, AA004644]

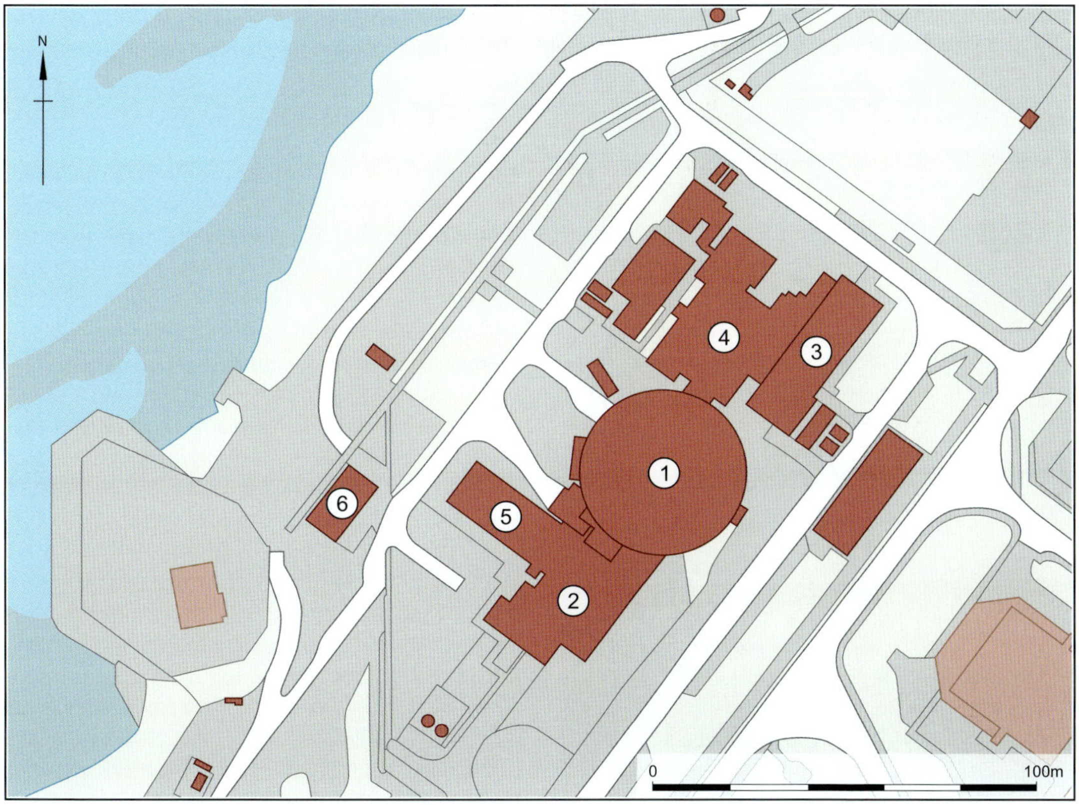

Fig. 4.22
UKAEA Dounreay, Caithness, fast-breeder reactor, showing principal components. Please note this 2024 plan also includes some additions built during decommissioning.

1 Reactor
2 Heat exchanger, link building and diesel generator
3 Administration building with control room
4 Active fuel-element storage pond and building
5 Turbine hall
6 Oil tanks (site of)

[Contains OS mapping. © Crown copyright (and database rights) 2025 OS AC0000815036]

The function of the spherical containment vessel was to contain any fission products in the event of an accident and, in the case of a coolant fire, to resist the resulting pressure differences. A sphere was the most efficient form as it offered the best volume-to-surface area ratio to house both the reactor and associated plant. This in turn determined that it should be 41m in diameter. To manufacture the sphere, the Motherwell Bridge & Engineering Co. Ltd set up their main office and works in one of the former wartime hangars in which an overhead crane was installed for moving the preformed plates. These were delivered to site singly, where they were carefully welded into pairs before being moved to the sphere for final assembly (Figs 4.23 and 4.24). In total, its mild steel plates weighed about 1,524 tonnes and are between 25.4mm and 47.6mm in thickness (Fig. 4.24). These were welded together to exacting standards, checked using X-rays to detect any cracks, and when complete the whole was pressure and vacuum tested. Entry into the sphere, both for personnel and rail-mounted bogies, was through an airlock accessed from the adjacent fuel element storage building. For the construction team the north of Scotland was particularly challenging, with short winter days requiring six-day working weeks to maintain the schedule. To avoid welding in freezing conditions, which might affect its strength, fabrication of the sphere took place in two campaigns in July 1955 and the following summer. Construction of the 9,144.42-tonne concrete reactor vault began in January 1956. The building site was covered with a canvas roof to reduce delays and provided with floodlights and steam heating.[65]

At the base of the sphere the reactor and associated primary coolant circuits and heat

Fig. 4.23
UKAEA Dounreay, Caithness, fast-breeder reactor, 1955. A pair of containment vessel plates are lifted into place in preparation for welding. [Courtesy of the Nuclear Decommissioning Authority, 8780]

Fig. 4.24
UKAEA Dounreay, Caithness, fast-breeder reactor, January 1957. The reactor vessel being lowered into place. [Courtesy of the Nuclear Decommissioning Authority, 0055 00073 AA004982]

exchangers were housed in the concrete biological shield mentioned above, known as the 'vault' (Fig. 4.25). This is bowl shaped, 27.43m in diameter, 13.71m tall, with 1.52m-thick walls. A hole in the roof slab provided access to the top of the reactor vessel. On the top of the vault wall was a single circular track for a rail-mounted rotating Goliath crane with two grabs rated at 25.4 tonnes and 7.62 tonnes. Above the reactor vault was a suspended, or false, steel floor, which formed the regular working level.

The reactor core was a hexagonal prism 0.53m across and 0.53m tall which comprised 345 annular fuel elements canned in an outer tube of niobium and an inner tube of vanadium. The reactor was controlled by raising and lowering 120 fuel elements into the core which held 225 fixed elements. The moveable rods were positioned at the base of the reactor and raised into the core by electromagnets, which in the event of a power failure would result in the rods dropping to their safety position. As a further safety feature, three neutron-absorbing boron rods could be inserted into the core. The core, breeder blanket and control elements were contained in a stainless steel cylindrical pressure vessel 3.2m in diameter and 6.32m tall which weighed 45.72 tonnes. This was constructed to very exacting tolerances by John Thompson Water Tube Boilers Ltd, Dudley, West Midlands. The quality of the welds was checked by over 4,000 X-rays.[66] This was surrounded by a 1.22m-thick borated graphite shield with a band of pure graphite at its equatorial position that permitted fission rate measurements to be taken. The core was surrounded by a blanket of 'fertile' material that reflected neutrons back into the core producing fresh fissile material. Each fuel element contained at either end a length of natural uranium which acted as the blanket or breeder material. The blanket was completed by surrounding the central portion with about 2,000 natural U-238 rods 2.44m in length and 3.18cm in diameter.[67]

The core was cooled by an alloy of sodium and potassium which became radioactive; this was circulated using electromagnetic pumps. Heat from the core was transferred through a double pipe to a secondary circuit, also containing sodium and potassium, which in turn passed through the biological shield and containment vessel to the heat exchanger house where it was used to produce steam for use in the adjacent turbine house.[68] The DFR achieved criticality in November 1959. It was primarily a research reactor; any electricity generated was either used by the establishment or exported initially solely to the local grid.

Periodically, the fuel elements in the core required replacing. To achieve this the rotating Goliath crane was used to manoeuvre the charge machine over two eccentrically rotating shields on top of the reactor vessel which allowed access to any position on top of the core. Irradiated fuel elements were raised into the charge machine which was heavily shielded and equipped to remove any residual heat from the fuel element. The crane then raised the charge machine and moved it over the canning station at the side vault. From there the fuel element was placed in a cooling pond for 120 days prior to reprocessing. Fresh fuel elements were held in the fuel element storage building and when required were placed into stainless steel cans filled with an inert gas before being loaded into the charge machine.[69]

The DFR successfully demonstrated that it was possible to safely operate a fast reactor cooled with a liquid metal coolant, and in 1962 was the first fast reactor to supply electricity to a national grid. Through operating this reactor, and research elsewhere, it became evident that the future Prototype Fast Reactor (PFR) would, for example, require different types of fuel and

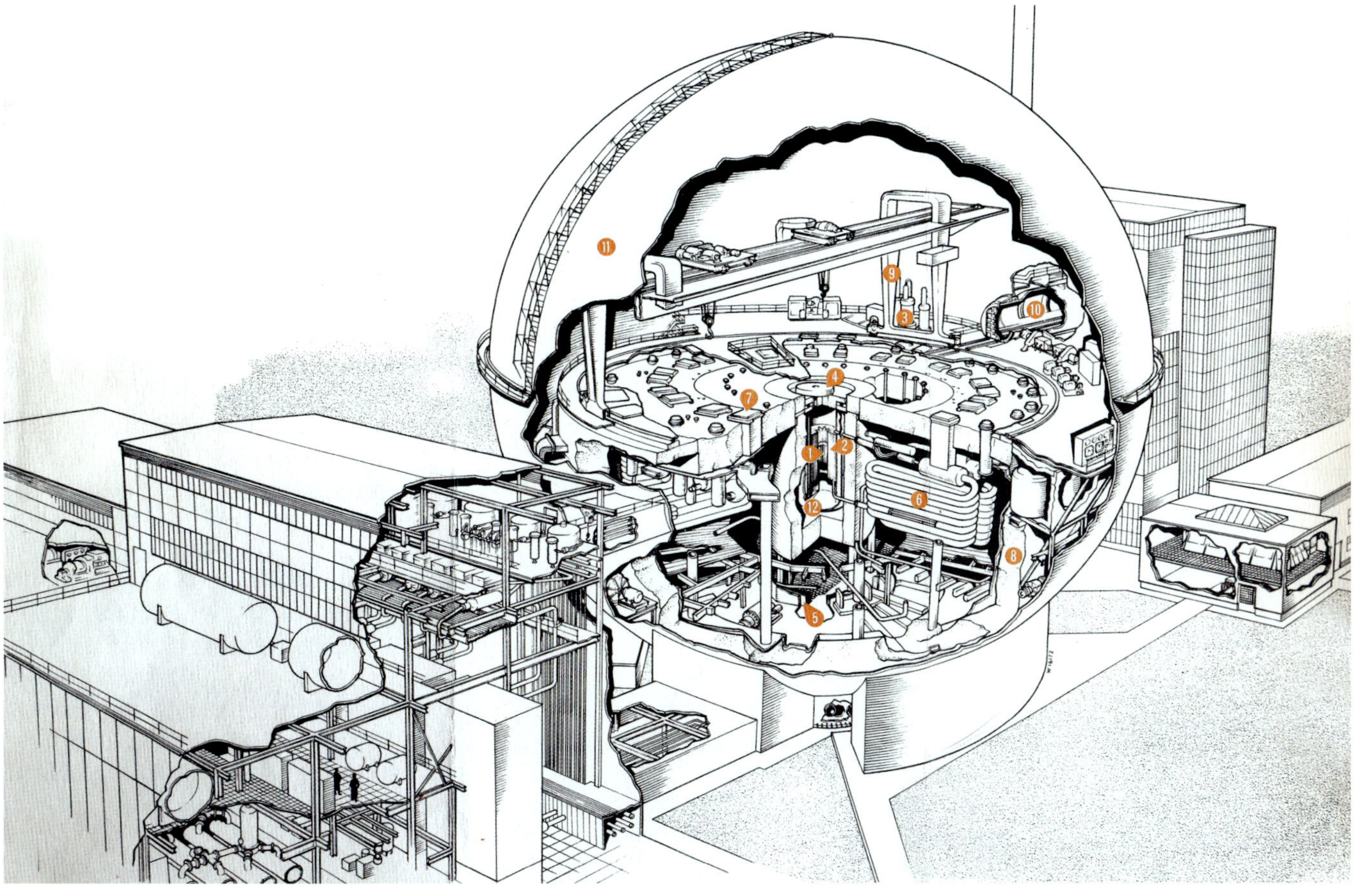

coolant circuits. The DFR therefore became less useful as a prototype for a full-size commercial reactor, and from the late 1960s to its closure in 1977 the focus of its work moved to materials testing, including commercial work for international clients.[70]

Dounreay Materials Testing Reactor (DMTR)

To support the DFR project and future reactor designs, the Dounreay Materials Testing Reactor (DMTR), similar to DIDO at Harwell, was installed and achieved criticality on 24 May 1958 (Fig. 4.26). It differed slightly in design from the Harwell examples and was configured to test larger engineering rigs. Experiments concentrated on the effects of radiation on construction materials, fuels and their canning materials, coolants and heat transfer media. The high neutron flux within the reactor considerably reduced the amount of exposure time items required compared to other reactors.

The reactor building is a cylindrical steel containment structure, 21.34m tall and the same in diameter. To prevent the escape of any radioactive particles, the building was entered through an airlock and held at a slightly negative internal pressure. Internally, there was enough space for the control room and manoeuvring trial rigs in the reactor. Its enriched uranium core, measuring 0.61m tall by 0.66m and 0.91m wide, was immersed in heavy water that acted as the moderator, coolant and reflector. This was contained in a 1.27cm-thick, 2m-diameter and 3.36m-tall aluminium tank, in turn surrounded by a graphite shield enclosed within a steel tank with a lead gamma shield, and the whole was surrounded by a 1.52m-thick concrete biological shield creating a cube with 6.1m sides. To conduct experiments, eight vertical tubes passed into the heavy water and six into the graphite reflector, and four horizontal tubes passed close to the reactor core.[71] The reactor ran until May 1969, when materials testing was concentrated on the DFR and the facilities at Harwell.[72] The control desk from this reactor has been preserved and is on display in the North Coast Visitor Centre, Thurso. At the time of writing, the containment structure had largely been stripped prior to demolition.

Fig. 4.25
UKAEA Dounreay, Caithness, fast-breeder reactor, cutaway diagram.

Key
1 Reactor core
2 Breeder blanket
3 Two charge/discharge machines and one transit coffin
4 Rotating top shields
5 Neutron shield
6 Primary heat exchanger
7 Electromagnetic pump
8 Biological shield
9 Rotating crane
10 Personnel air lock
11 Containment vessel
12 Reactor vessel

[Courtesy of the Nuclear Decommissioning Authority]

The Chemical Group

Given the remoteness of Dounreay, it was decided to make the facility self-sufficient by providing it with a dedicated Chemical Group, later known as the Fuel Cycle Area (FCA). This was situated to the east of the establishment and was divided into four main sections: fuel element fabrication, chemical separation and billet production, effluent disposal, and chemical analysis. The fuel element production lines were divided into two, one for the fast reactor and the other for the materials testing reactors at Dounreay and Harwell, and British Commonwealth reactors of this type. To recover the uranium from the irradiated fuel it was dissolved in nitric acid before solvents were used to separate the uranium, fission products and plutonium. Within the plant, most processing was carried out behind 1.22m-thick concrete walls until the radiation levels had been sufficiently lowered to permit handling in glove boxes (Figs 4.27 and 4.28). The recovered uranium was then rendered into billets and returned to the fuel element production line. High-level waste from these processes was stored in underground tanks, while low-level effluent waste was discharged into the sea. Quality control throughout the production and solvent recovery processes was overseen by the chemical analysis laboratory.[73]

It was characteristic of research establishments that new requirements led to additional facilities. In 1956, to support research into fast reactor fuels, the small ZETR research reactor was moved from Harwell to test cell 1 in building D1249 at Dounreay. It had the distinction of achieving the first criticality in Scotland on 13 August 1957 and operated until early 1961.[74] In anticipation of reactor operations with associated fuel preparation and reprocessing, dedicated facilities were constructed to understand the safety of solutions containing uranium and plutonium, and the prevention of criticality incidents in processing plants, laboratories, fissile material stores and other places (Table 4.2).[75] Complementing earlier work at Harwell and ongoing research at Aldermaston, a purpose-built Criticality Laboratory D8550 was sited to the east of the original Chemical Group. It only operated for a few years, until 1963 when it was partly dismantled and mothballed. Further additions were made to the Chemical Group including, in the early 1960s, new plutonium examination facilities with three suites of heavily shielded caves for inspection and reprocessing.[76] In 1984 a cementation plant was installed for mixing liquid waste with cement.[77]

Fig. 4.26
UKAEA Dounreay, Caithness, materials testing reactor, showing its containment vessel. In front, the office building with it steel frame and light cladding is typical of many structures within the nuclear research establishments.
[Courtesy of the Nuclear Decommissioning Authority, 16774 A011347]

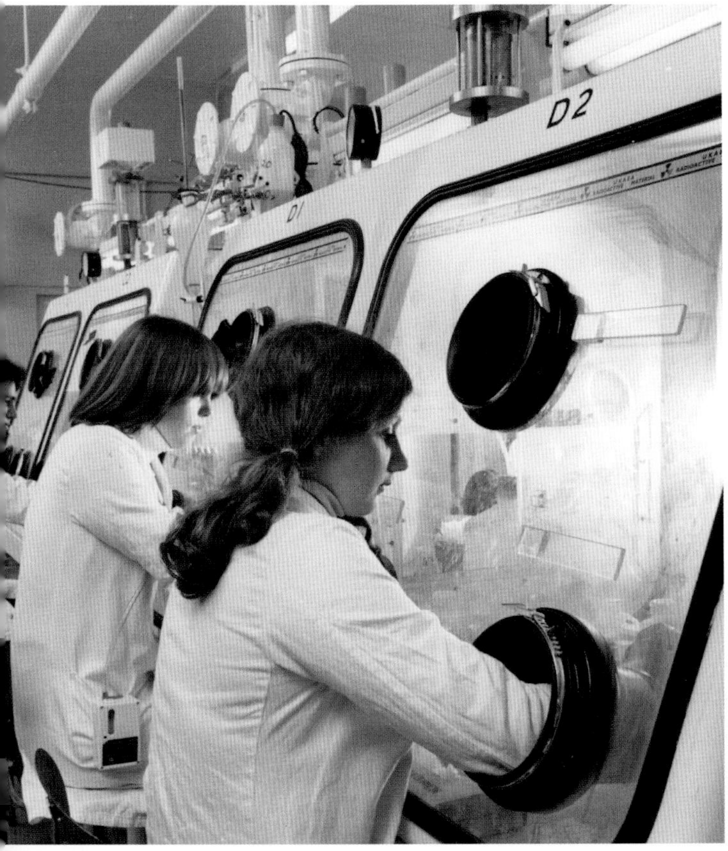

Table 4.2 Dounreay reactors and critical facilities.

Name	*Operational*	*Notes*
ZETR 1b	1957–60	Modified ZETR 1a (Harwell), studies of uranium solutions
Materials testing reactor	1958–69	
Fast reactor	1962–77	
TESSIE	1959	Safety of sludges and precipitates
PHOENIX (HAZEL II)	1960	Safety of solutions in process vessels
PUMA	1960	Study of under-moderated Pu/H systems
SIRIUS	1961	Neutron shielding studies
TOAD – ZETR III	1961	Solution safety in small vessels
PANTHER	1963	Safety of plutonium solutions
QUAGGAR	1963	Dilute fast-reactor physics
Prototype Fast Reactor	1974–94	

Main source: European American Committee on Reactor Physics 1964 *List of Research Reactors and Critical Facilities.*

Fig. 4.27
UKAEA Dounreay, Caithness. Personnel testing samples from the reprocessing plant in glove boxes in the medium-active plutonium laboratory. [Courtesy of the Nuclear Decommissioning Authority, AJ8288 C]

Fig. 4.28
UKAEA Dounreay, Caithness, high-activity analytical laboratory, 1979. The suite of 11 interconnected shield cells was used for the analysis of highly radioactive samples from the chemical plant using master–slave manipulators. [Courtesy of the Nuclear Decommissioning Authority, AJ8284/3 C 52/249 (B)]

Box 4.3 Nuclear propulsion

In the 1950s, the only constraints to the application of nuclear energy for propulsion on land and sea, and in air and space, appeared to be the imaginations of the nuclear scientists.

Nuclear power at sea

For seafarers, the use of nuclear power potentially offered possibilities of greater speed and unlimited cruising ranges unhindered by the necessity to dock at coaling stations or oil depots, the principal limitations being the endurance of the crew and food supplies. In a typical marine system, a nuclear reactor is employed to produce high-pressure steam, which is used to turn a turbine to generate electricity to power a vessel's electrical systems and propulsion motor.

Over the last 60 years, the Royal Navy has been the country's largest operator of nuclear cores, with around 80 in service at sea and on shore.[1] The Admiralty scientific staff for the Royal Navy had acquired early knowledge of the possibilities of atomic energy through sometimes reluctant secondments to the wartime Tube Alloys Project. Post war, from the beginning of the Atomic Energy Research Establishment (AERE) at Harwell, a small Admiralty contingent was present to keep up to date on developments that might present opportunities for naval engineering.[2] As early as the late 1940s, discussions began between AERE and the Metropolitan Vickers Electrical Company about a design study that might lead to the development of a submarine propulsion reactor. The conclusion of the research in May 1952 was that current gas-cooled reactor technology would be unsuitable for use within a submarine.[3] As a fuel it would also require an assured supply of enriched uranium that was, at that time, uncertain. After this initial setback, the launch of the United States' nuclear-powered submarine USS *Nautilus* in January 1954 pointed the way to the use of compact pressurised water reactors. The same year, the Harwell naval section began to expand, and attention was turned to investigating how to shield a reactor within the confines of a submarine.[4] To support this naval work, a swimming pool-type reactor, LIDO, was built on the new reactor site between PLUTO and DIDO (*see* Fig. 4.13). It comprised a core of enriched U-235 suspended in a Portland concrete tank 8.53m by 2.44m and 7.32m deep. This was filled with demineralised water that acted as the moderator, coolant and shield. On one side of the tank wall was an opening sealed by an aluminium panel. During trials, the core was manoeuvred against this panel and various test materials were placed against it to measure their effectiveness as shields.[5] Around the same time, Harwell, in collaboration with English Electric and Babcock and Wilcox, began design work on a pressurised water reactor, LEO, but it never progressed beyond the drawing board.[6] To investigate naval propulsion, the Admiralty also operated another small swimming pool-type reactor at Harwell, NEPTUNE, between January 1957 and June 1959; it was later transferred to Rolls-Royce's Raynesway factory in Derby.[7]

With preliminary reactor work underway at Harwell in January 1956, the Treasury agreed to fund the construction of Britain's first nuclear submarine and an associated shore prototype and training establishment. After considering Windscale and Dounreay, the latter was selected, partly due to the greater availability of space. Construction began in 1957 with the main plant and machinery supplied by Vickers-Armstrongs (Engineers) Ltd and the reactor by Rolls-Royce. Initially, it was known as the Admiralty Reactor Test Establishment Dounreay, and from the mid-1960s was managed on the Admiralty's behalf by Rolls-Royce. In 1970 it was commissioned as HMS *Vulcan* and later known as the Vulcan Naval Reactor Test Establishment.

It was adjacent to, but separate from, the Dounreay Experimental Research Establishment (DERE) site, which supplied transport, housing, electrical, waste, and emergency services to the Admiralty. DERE also took responsibility for the new establishment's civil engineering and building design. Its principal structures comprised a large building to house the Dounreay Submarine Prototype 1 (DSMP-1), an administrative block, classrooms, workshops, ancillary services and a pumphouse. Accommodation was provided in the purpose-built Scapa House, Castlegreen Road, Thurso (now demolished). The prototype was effectively the aft end of a Valiant-class submarine hull made up of a reactor and drive train. Its original function was to demonstrate the machinery would fit inside a submarine hull, and was later used for engineering tests and crew training. It operated from 1965 to 1987.[8]

In the late 1950s within the DERE site, probably in one of the wartime hangars, work also began to refine the layout of equipment bays via a wooden mock-up of a submarine reactor and associated machinery spaces; this was subsequently moved to the Vickers Supermarine factory in Southampton.[9] Meanwhile, in 1958, the conclusion of the Mutual Defence Agreement with the United States gave the United Kingdom access to pressurised water reactor technology suitable for use in submarines. Through direct technology transfer, Britain's first nuclear submarine, HMS *Dreadnought*, was powered by a Westinghouse S5W reactor. A condition of the assignment laid down by United States Admiral Rickover was that it should be a company-to-company handover, namely from Westinghouse to Rolls Royce.[10]

To assist in the design and performance studies of the second-generation reactors, the Shore Test Facility (STF) was opened in 1986 at Dounreay, which ran until 2015. Accumulated experience from the operational fleet, advances in computer engineering and computer training simulators required a less elaborate prototype, and this facility just comprised a representative core and pressure vessel.[11] For the current third-generation reactors, such is the confidence in computer modelling, modern materials and manufacturing quality assurance, no requirement is envisaged for a full physical prototype.

Submarine reactors are manufactured at Rolls Royce's Raynesway factory, Derby, where it also maintains the NEPTUNE reactor test facilities. In anticipation of the PWR3 programme, the original 1960s manufacturing plant has recently undergone a £500 million refurbishment programme.[12]

To service the new fleet, a purpose-built Nuclear Submarine Refitting Facility was built at the eastern end of Chatham dockyard (Medway) alongside No. 1 Basin, opening in June 1968. The main contractor for the work was John Mowlem and Company Limited. At the head of the dock was a ten-storey accommodation block, which provided housing for the submarine crews, a canteen and office space, which together allowed for continuous work on the boats. Adjacent to this was a two-storey combined trades workshop. Dominating the facility was a

Fig. B4.3.1 Pembroke Dock, Chatham, Medway, Nuclear Submarine Refitting Facility, October 1968. To the left was the control room and adjacent to it the tall building housed the core pond. Dominating the facility was the 49m refuelling crane, and to its right were workshops and the office block, with crew accommodation and a canteen. [© Historic England Archive, PG 12824_012]

48.76m-tall, 1,524-tonne, hammer-head crane, which served the dry docks to either side (Fig. B4.3.1). Facilities to handle radioactive components were housed in a single structure and included a wash-down area, effluent treatment plant, core pond, refuelling equipment shop and training rig, and 'health physics', which was a team who were responsible for monitoring personnel and equipment, and if necessary, decontamination. During refuelling, a special reactor access house was lowered into position onto the submarine's hull; the spent reactor was then lifted clear and placed in the core storage pond. The first refit was completed in June 1972 and the final one in May 1983. The facility was subsequently demolished, and its site is now covered by retail and residential property. Today, the main operational focus for the ballistic missile submarine fleet is in Scotland, with its principal base at Faslane, an armament depot at Coulport (Argyll and Bute), and maintenance and refit facilities at Rosyth (Fife) shipyard.

From the use of reactors in submarines it was a short move to apply this technology to surface shipping. The United States has been the most enthusiastic adopter of this technology, with 20 nuclear-propelled naval vessels, while the Soviet Union launched 5 nuclear-powered naval vessels. Beyond the Cold War superpowers, only France has operated a nuclear-powered warship.[13]

The first nuclear-powered commercial ship was the Soviet icebreaker, *Lenin*, which entered service in 1959; Atomflot continues to operate the latest generation of nuclear-powered icebreakers from Murmansk. In 1962 the United States launched the research ship the NS *Savannah*, powered by a pressurised water reactor. The only European venture in this area was the West German freighter and research vessel the *Otto Hahn*, named after the physicist. This was viewed as a means of promoting the German shipping and emerging nuclear technology industries. Ultimately, it was non-technological factors, including the availability of docking licences, insurance and operational costs, that put paid to this project. Furthermore, pressure had been put on Brazil, a potential development partner, to sign the Nuclear Non-Proliferation Treaty and withdraw from the development of nuclear technology.[14] The only other nuclear-powered ships were the Japanese *Mutsu*, and the Soviet *Sevmorput*, which remains in service. From the late 1950s, the UKAEA initiated several studies into nuclear-powered ships and there were frequent questions in Parliament about Britain lagging behind in this field. The practicalities and economics of operating purely commercial nuclear-powered ships forestalled plans to develop this technology.

Nuclear-powered flight

The long duration between refuelling made nuclear power attractive to aircraft designers, and in 1946 the United States Army Air Force began research in this area.[15] In such a system, instead of burning liquid fuels, compressed air within a jet engine would be heated by a reactor. In the 1950s, it was envisaged that a bomber powered by a reactor might represent the ultimate airborne deterrent, remaining aloft as long as supplies for the crew lasted. The main drawbacks to the use of reactors were their weight, shielding needed to protect the crew, and the hazards that would result from a crash.

The sustained power that could be delivered by a nuclear-powered engine also opened possibilities for extended space missions. In such a scenario a reactor might be used to heat liquid fuel into gaseous form before it was expelled through a nozzle. A second option might be to use a reactor to generate electricity and to positively charge a propellent and to push the resulting ions out through a thruster.[16] In 1958, the Nuclear Rocket Development Station was established within the Nevada Test Site. The facilities included trials sites linked by rail to the heavily shielded R-MAD building for reactor maintenance, assembly and disassembly. This staged programme progressed from the Rover research reactor to the flightless Kiwi series, and on to NERVA (nuclear engines for rocket vehicle application). The flight stage was cancelled in 1963, and the whole programme terminated in 1973.[17] Elsewhere, within the range, the University of California Radiation Laboratory established facilities for testing PLUTO, a nuclear-powered ramjet for missile propulsion. It was equipped with similar facilities to the rocket programme but was cancelled in 1964. Over the past two decades, the Desert Research Institute, Las Vegas (Nevada), has been actively researching these largely forgotten relics of abandoned nuclear space technology.[18]

On land, atomic railway locomotives might be able to support the weight of a reactor and its shielding but would be vastly more expensive than other power sources, and vulnerable to hijack.

Dounreay Prototype Fast Reactor

In a fast reactor, enriched fuel, U-235 or plutonium, is surrounded by a fertile fuel, U-238, which is converted to fissile fuel by bombardment with neutrons escaping from the core, resulting in a net gain of fissile material. Initially, enriched uranium was used as the fuel, but there was a contemporary expectation that by the mid-1960s there would be a surplus of plutonium from the thermal stations that might be used to fuel a new generation of fast-breeder reactors.[78] It was envisaged that the future fast breeders would form part of a balanced national nuclear reactor fleet system mainly fuelled by plutonium recovered from the thermal reactors. They were seen as the reactors of the 'next century' and the most efficient form of fission reactor.[79] Eventually, as the reactors bred more enriched fuel than they consumed, it was believed they might allow the United Kingdom to operate almost independently of international uranium supplies.

The choice of Dounreay for the site of the Prototype Fast Reactor (PFR) was announced to the House of Commons in February 1966. Other locations had been considered at Chapelcross (Dumfries and Galloway), Windscale (Cumbria) and Winfrith (Dorset); in addition to the technical issues, the decision was also swayed by regional and social considerations, with strong local lobbying for the project to be located at Dounreay. The estimated cost of the scheme was £30 million, and it was expected that the reactor would be operational by 1971. The cost included £1 million allocated to the construction of an associated fuel production plant at Windscale and included the cost of linking 275kV transmission lines to the Scottish grid. The electrical output was to be 250MW, a quarter of the output foreseen for the future commercial stations planned for the late 1970s. The PFR was owned and operated by UKAEA. However, in order to gain the maximum benefit from this prototype, responsibility for its construction was passed to the Nuclear Power Group, with design offices at

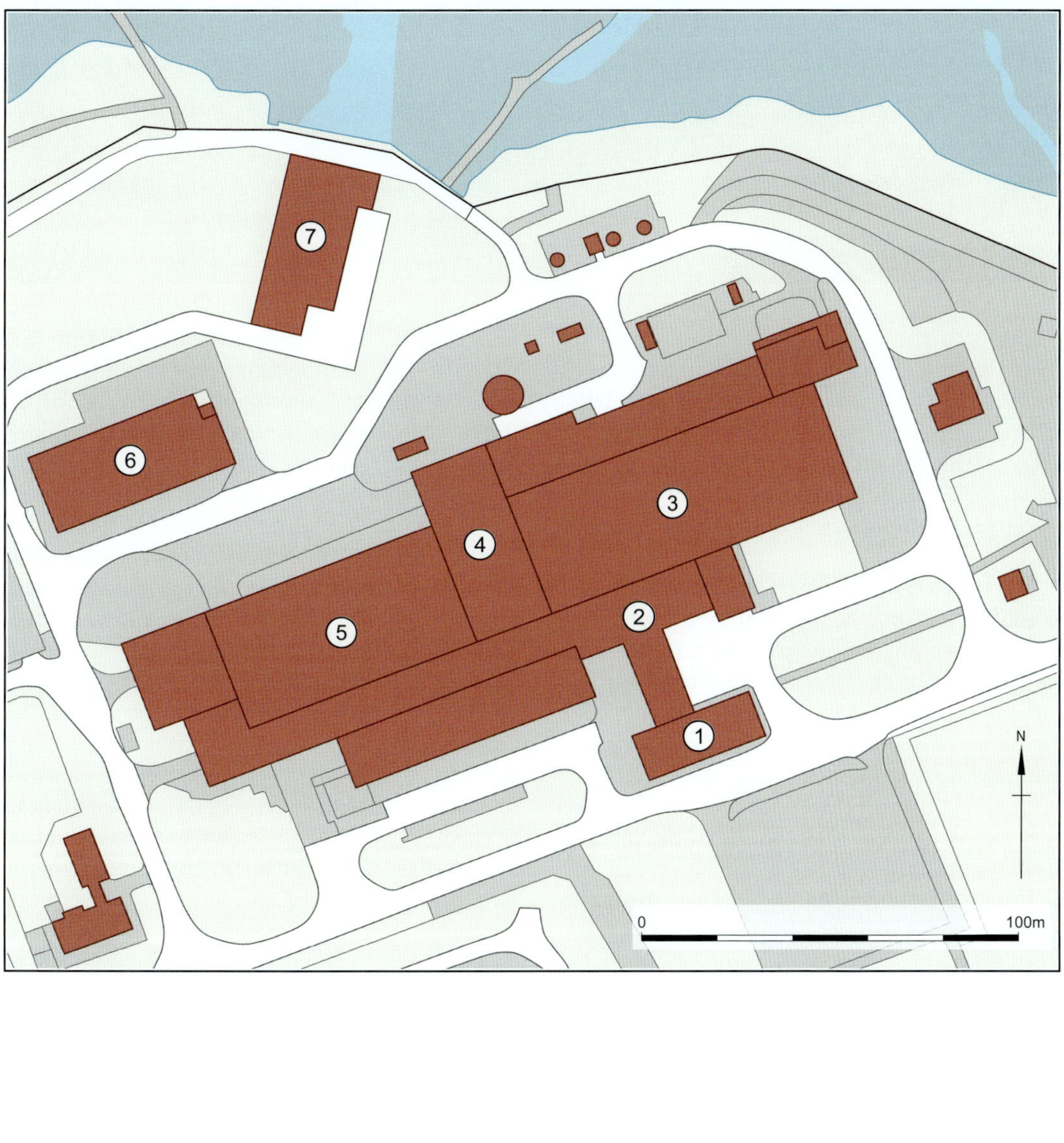

Fig. 4.29
UKAEA Dounreay, Caithness, Prototype Fast Reactor.
Key
1 Administration building
2 Control room
3 Reactor building
4 Steam generator building
5 Turbogenerator building
6 Sodium store
7 Seawater pump-house

[Contains OS mapping. © Crown copyright (and database rights) 2025 OS AC0000815036]

Fig. 4.30
UKAEA Dounreay, Caithness, Prototype Fast Reactor, June 1989. To the rear is the white dome of the fast reactor and beyond, the white containment vessel of the materials testing reactor. [Courtesy of the Nuclear Decommissioning Authority, 7050-0774]

Risley, with the intention that this experience might be transferred to later commercial fast reactors. In the event, construction was beset with supply and technical issues as well as labour strife, resulting in a three-year delay to the project. The PFR finally began to run on low power on 2 March 1974 and was fully operational by 1975.[80]

The PFR was sited to the west of the original establishment on a slightly different alignment, the main axis aligned roughly west to east, with the Vulcan Naval Reactor Test Establishment immediately to its west. To the north, between the main PFR building and the sea, were the freestanding sodium store and seawater pumphouse and to the south was the covered electrical switchgear that connected the plant to the National Grid (Fig. 4.29).

The anonymous main building measures 204m in length with a maximum width of 96m. It is steel framed, with the varying rooflines and cladding materials providing the only outward expression of the functions within; most external doorways are painted yellow (Fig. 4.30). To move plant components within the building, most of the separate building frames carried travelling cranes. To the south is the three-storey administration building, connected to the main building by the link building that provided access to the control block. The main reactor hall was to the east, clad externally with concrete panels. To create a sealed structure these were bonded together and to the frame by polysulphide. This effectively formed the secondary containment for the reactor within the reactor hall, designed to resist any changes of pressure or releases of radioactivity due to an accident in the reactor or during refuelling. The ventilation system was also engineered to maintain negative pressure in the building to prevent air egress.[81]

The reactor was positioned at the west end of

this building, and to reduce the amount of shielding required was sunk into a below-ground vault (Fig. 4.31). To its east were shielded caves for handling irradiated fuel elements where, in keeping with the experimental nature of this plant, fuel elements might be partly disassembled and examined.[82] The reactor core was 1.67m in diameter and 0.91m tall. The primary circuit, comprising the core, intermediate heat exchangers and fuel-handling system, was contained within a 12.19m-diameter, 15.2m-deep, stainless steel tank. This is in turn surrounded by a leak jacket to prevent the loss of the sodium coolant, 1,016 tonnes of which was contained in the primary circuit and 203.2 tonnes in the secondary. In contrast to the DFR, the PFR was a so-called 'pool' type whereby the reactor, full primary coolant circuit, coolant pumps and intermediate heat exchangers were immersed in the sodium contained by the reactor vessel.[83] To reduce stresses on the vessel and prevent the core from being exposed, no openings were made below its top; all connections were made from above and all the main components were suspended from the roof.[84]

The reactor was fuelled by a mixture of 4.06 tonnes of plutonium and uranium oxide pellets clad in a stainless steel alloy. The fuel assemblies were formed of a hexagonal frame 3.81m in length within which were 325 fuel rods with an outer diameter of 5.84mm and 2.25m in length: the active length holding the plutonium was 0.91m. Above and below this were pellets of depleted uranium, and above the top blanket 19 shorter rods with a diameter of 19mm. The core was assembled into a complex honeycomb pattern of fuel assemblies, control rods, and experimental tubes (Fig. 4.32). Towards the centre of the core were plutonium and uranium oxide assemblies, the proportion of each depending on their position in the core. These were surrounded by breeder assemblies and reflector assemblies.[85] Given the reactor's experimental nature, fuel loading might vary to accommodate experimental needs.[86] The reactor was controlled by five tantalum control rods and five boron carbide shut-off rods.

While shut down the reactor was charged by a machine suspended from a rotating shield above the core that provided access to all the fuel positions. The fuel pins were arranged in hexagonal clusters and on removal were held for 30 days in a transfer rotor, submerged in the primary sodium coolant before removal to external fuel-handling caves to be prepared for transport to the reprocessing plant (Fig. 4.33).[87]

In the centre of the complex was the Steam Generator Building, its projecting upper section clad in black glass. Here heat was transferred from the secondary coolant system to the steam generators to power the turbine. Black glass cladding was also used on the tall block to the north housing the blanket gas plant, ventilation plant and associated switchgear. At the western end of the main building stood the Turbine Hall, lit by lantern-style roof lights. To lessen the consequences of a catastrophic turbine failure, the shaft was aligned with the building, keeping the

Key
1 Reactor core
2 Intermediate heat exchange (6)
3 Primary sodium pump
4 New and irradiated fuel transfer flask
5 Fuel transfer system
6 Fuel examination facilities
7 Fuel transit flask
8 Fuel storage
9 Experimental equipment
10 New fuel flask
11 Secondary sodium pump (3)
12 Sodium water reaction dump tanks
13 Steam drum (3)
14 Expansion tank (3)
15 Superheater (3)
16 Evaporator (3)
17 Reheater (3)
18 Circulating pump (3)
19 Turbogenerator
20 Reactor operating floor
21 Wet area
22 Thermal syphon system
23 Steam generator building
24 Reactor building
25 Turbogenerator building
26 Primary cold-trap loop
27 Station control room
28 Vehicle airlock
29 Administration building
30 Electrical annexe
31 Diesel generators
32 Transformer to grid

[Courtesy of the Nuclear Decommissioning Authority]

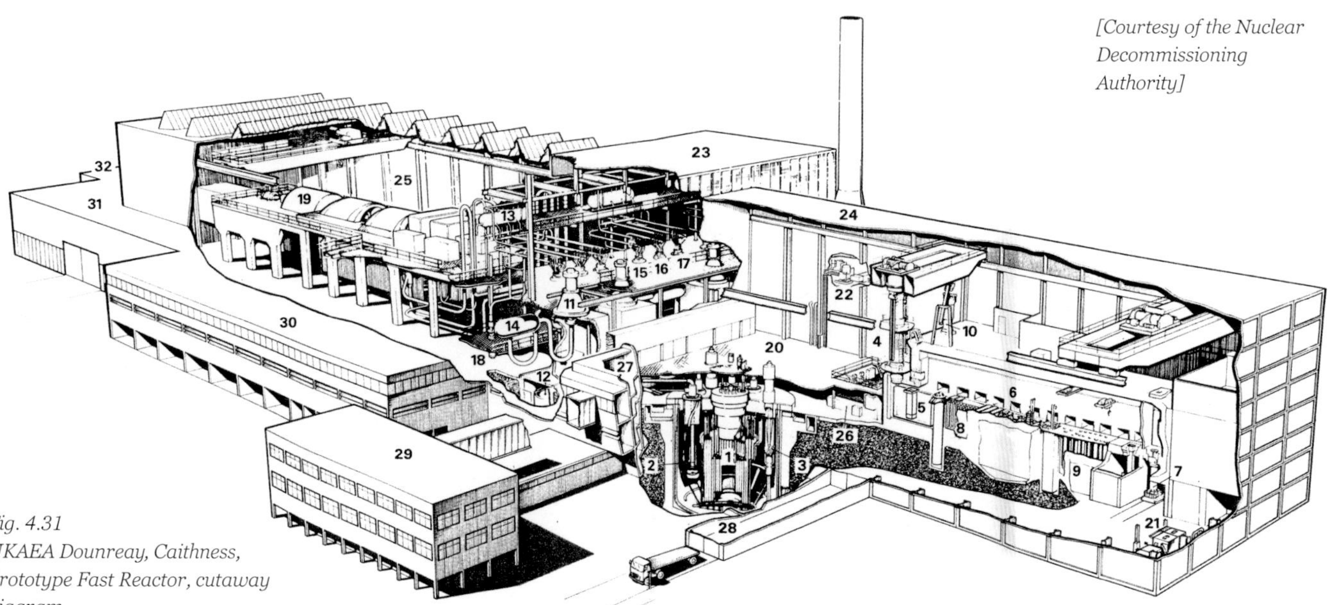

Fig. 4.31
UKAEA Dounreay, Caithness, Prototype Fast Reactor, cutaway diagram.

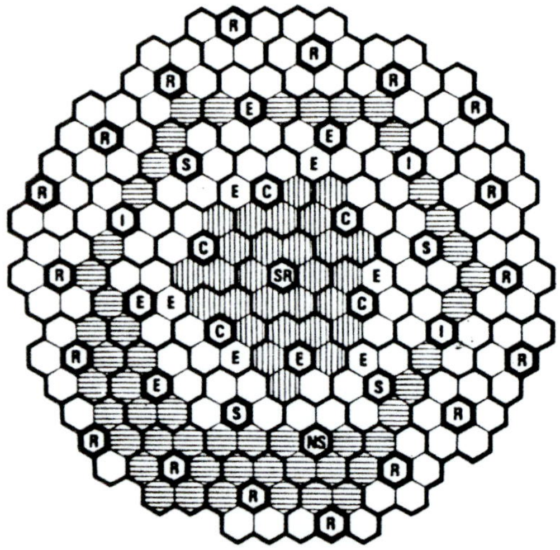

Fig. 4.32
UKAEA Dounreay, Caithness, Prototype Fast Reactor, diagram of the honeycomb arrangement in the core.
C – Control rod
S – Shut-off rod
SR – Safety rod
NS – Neutron source
R – 4-inch reflector
I – Instrument

[Courtesy of the Nuclear Decommissioning Authority]

reactor core out of the path of any jettisoned fragments. Attached to the west wall of the main building were the diesel generator building and its transformer. The PFR operated until 31 March 1994.

To reprocess the fuel from the PFR, in 1971 it was decided to decontaminate and rebuild the reprocessing plant that had served the original fast reactor. Once in operation, plutonium would be separated out as a plutonium nitrate solution and transported by sea to Sellafield for further reprocessing and incorporation into new fuel elements. The rebuilt plant was opened by Margaret Thatcher in September 1979.[88]

Fig. 4.33
UKAEA Dounreay, Caithness, Prototype Fast Reactor, October 1974, reactor hall.
[Courtesy of the Nuclear Decommissioning Authority, AJ7734 C AA010018]

Later years

Through to the late 1970s fast-reactor technology continued to be viewed as an attractive option for the future, and in the early 1980s a collaborative European project was established. In this it was envisaged that Dounreay would play a key role reprocessing the fuel from all the stations across the Continent. In preparation for this, the Marshall Laboratory was opened in 1986, named after Walter Marshall, then chairman of the Central Electricity Generating Board, to support the design of the European Demonstration Reprocessing Plant (EDRP).[89]

In 1988, the government announced that it intended to cease funding fast-reactor technology; the EDRP project was abandoned and the PFR shut down in 1994. A further setback occurred in 1998 when a fault in one of the fuel reprocessing dissolvers led to the end of this activity.[90] Subsequently, the site's primary activities have been decommissioning the legacy of over 40 years of cutting-edge nuclear research. This activity has involved multi-million-pound investments to build new facilities to treat irradiated coolants and intermediate waste, low-level waste disposal facilities, and storage vaults. To reduce the volume of stored waste, the Waste Receipt Assay Characterisation and Supercompaction facility has also been added.

Atomic Energy Establishment, Winfrith, Dorset

By the early 1950s Harwell was becoming crowded, with few available building sites or capacity within its service sections and infrastructure to support new developments. There were also concerns about accommodating an ever-expanding workforce and causing any increase to the amount of effluent being discharged into the Thames. In particular, more space was required for the UKAEA's Reactor Group to build large prototype reactors. The search for a new site had begun by summer 1954; location criteria included ready access to Harwell, relative remoteness, a reliable water supply for cooling, and a means of disposing of liquid effluent. By spring 1956, after more than 50 possible places had been considered, Winfrith Heath was chosen.[91] Planning permission was given in November of that year, and after a three-day planning inquiry in January 1957, consent to begin work was granted. Construction began on 2 September.[92]

The site selected lay immediately to the south of the Dorchester-to-Wareham railway line, midway between the two towns and nearly 1km west of the village of Wool. An area of 278ha was acquired for the new location, comprising a mixture of heath and agricultural land – the only 'greenfield site' ever acquired for a major atomic establishment. Knighton Heath, a few kilometres to the south-west, had been used as a wartime bombing decoy to protect RAF Warmwell (Dorset), and the latter's abandoned buildings provided accommodation for some of the construction team.[93] Responsibility for the design of the new establishment lay with the UKAEA Development and Engineering Group based at Risley, including the director of works and buildings, T. L. Viney; chief architect, R. S. Brocklesby; and resident engineer T. G. Williams. The main contractor for the construction work was the Turriff Construction Corporation Ltd.

The site layout followed the roughly north-west to south-east orientation of the Dorchester-to-Wool railway line, and its central complex of buildings covered an area of about 1,400m by 600m (Fig. 4.34). To give a campus-like quality, the main buildings were well spaced, with gaps providing room for future expansion. There were local concerns about the visual impact of the new establishment and the industrialisation of an otherwise rural area, and accordingly the county council advised that it should blend into the local countryside. In order to mesh the new scheme with its surroundings, formal planting was restricted, parts of the former fields were maintained as mown grass with remnant trees and hedge lines, and elsewhere the heathland cover was retained (Fig. 4.35). Buildings were generally kept below 15.24m and were faced in a mixture of buff and dark Swanage bricks. Where asbestos sheets were used, instead of their usual light-grey finish, they were coloured in two shades of grey/green. Roofing materials included copper sheet (which weathered to green), mineral felt and coloured asbestos.[94]

The central group of buildings provided a mixture of administrative and technical support activities, and also included halls for zero-power reactors and other experimental rigs (Fig. 4.36). One of the largest buildings, and one of the first to be finished, was the Reactor Group's offices (Fig. 4.37). This was located to the north of the

Fig. 4.34
AEE Winfrith, Dorset. Diagram showing the main features of the establishment. [Contains OS mapping. © Crown copyright (and database rights) 2025 OS AC0000815036]

Key
1 Main gate
2 DRAGON project office
3 Apprentice training school
4 Social club
5 General services building
6 Reactor physics building
7 Zero energy buildings
8 Canteen
9 Computer centre
10 Reactor Group offices
11 Medical centre
12 Reactor physics
13 Project technical building
14 Physics technical building
15 Plutonium fuel laboratory A52
16 Active handling building A59
17 Reactor school
18 Police lodge
19 Main stores and garage
20 DRAGON reactor facility
21 Steam-generating heavy-water reactor
22 ZEBRA reactor facility
23 Reactor safety compound

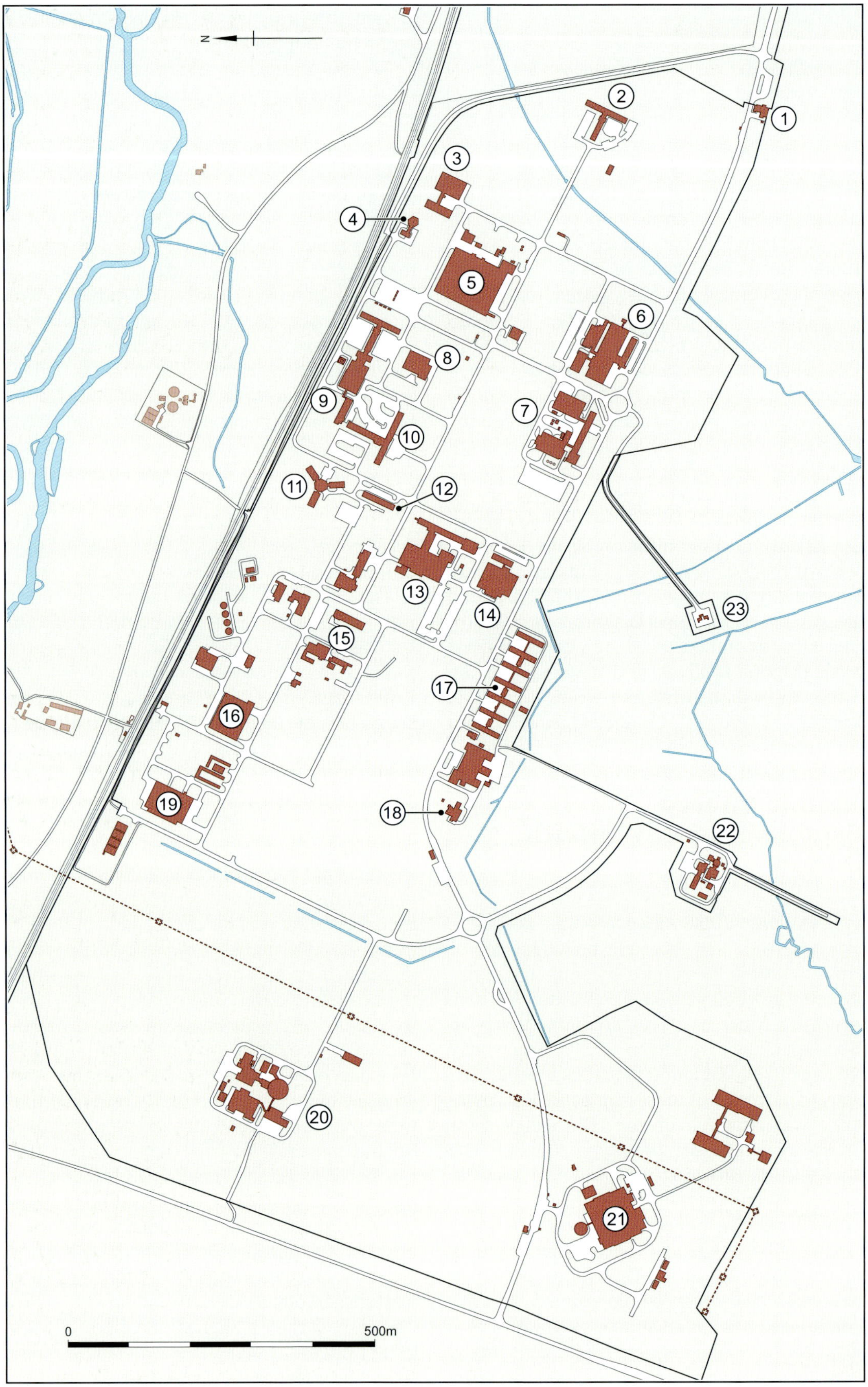

site and was five storeys high with a T-shaped plan; the east-to-west wing measured 92.35m by 9.14m and the north-to-south wing 64m by 9.14m. The main structure was of reinforced concrete, with the side panels infilled with buff and red bricks to create a chequerboard pattern with bespoke metal window frames. The top storey was of steel portal frame construction with a steel 'butterfly' roof drained by a central gutter and clad with light glazed walls; it was intended to be used for drawing offices and directors' meeting rooms.[95]

To the north, the Group Office was linked to the two-storey, concrete-framed Computer Building, 42.67m by 12.19m. Access to high-powered computers was essential to many studies. An Electronics Associates PACE 231R analogue computer was in operation by 1959, and in the following year a Ferranti Mercury digital computer was installed. Over the following years a succession of room-sized mainframe computers was introduced; the installation of

Fig. 4.35
AEE Winfrith, Dorset, showing how the establishment was laid out following the grain of the land defined by the railway; at ground level trees obscured its structures. Within the site boundary swathes of grass and heathland blend with the local landscape.
[© Historic England Archive, 26652_001]

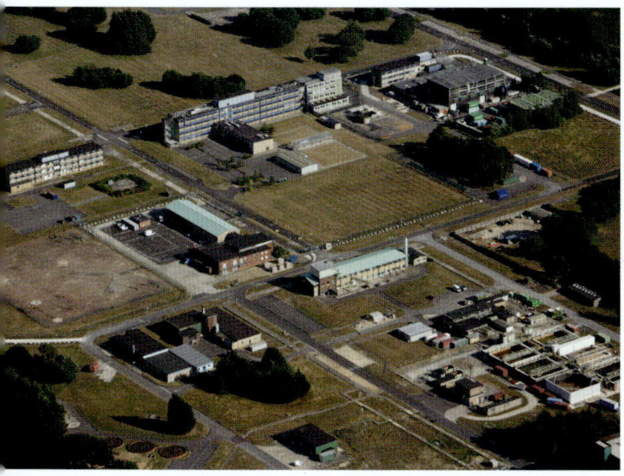

Fig. 4.36
AEE Winfrith, Dorset, central area showing administrative offices and workshops.
[© Historic England Archive, 26652_028]

Fig. 4.37
AEE Winfrith, Dorset, Reactor Group offices, showing the chequerboard effect created by different coloured bricks, and the upper storey that housed drawing offices and conference rooms.
[Courtesy of the Nuclear Decommissioning Authority]

Fig. 4.38
AEE Winfrith, Dorset, canteen. To the left the internal staircase to the director's dining room is visible and beyond the car port the entrance for other staff. On the roof was the aluminium-clad tank and plant room.
[Courtesy of the Nuclear Decommissioning Authority]

Fig. 4.39
AEE Winfrith, Dorset, social centre. This timber building with a copper-clad spire was constructed by Simms Sons & Cooke, Limited.
[Courtesy of the Nuclear Decommissioning Authority]

the ICL 4-70 in 1969 necessitated an extension to the building. Mainframe computers survived until 1989 when they were replaced by two VAX machines and local microcomputers. Remote access was also possible to the Cray 2 facility at Harwell.[96]

By 1960 the staff at Winfrith numbered 1,870; to cater for this workforce an impressive canteen was provided close to the Reactor Group offices.[97] It was a large, 41.45m by 32m, steel-framed, double-storey structure with a curved steel-decked roof, on top of which was a pear-shaped, aluminium-clad plant room and tank (Fig. 4.38). On the ground floor was a snack area with a grill, servery and bar. Ramps led up to the main dining area where food was collected from the servery, with the main kitchen to its rear. An alternative was a waitress-service dining room accessed from a separate staircase. In keeping with contemporary grading distinctions, there was a dedicated dining room for the directors and senior staff with its own distinct entrance.[98]

Staff welfare was also catered for on the northern side of the site by the hexagonal social centre set beneath a copper roof surmounted by a spire, mirrored internally by an open inverted spire providing central support (Fig. 4.39). It was designed by a local architect, Ron Snadden, and the structure was supplied by Simms Sons & Cooke Ltd.[99] Close by was a tennis court and bowling green, with playing fields adjacent to the main gate. Other supporting buildings included the entrance lodge, medical and health physics facilities, stores, a garage, substations, a telephone exchange, and ambulance and fire stations.

Manufacturing

Many of the reactor experiments required special types of fuel, either to trial new fuel combinations, the sorts of material used to can the fuel rods and their forms, or to model the changing proportions of elements found in reactor fuels when subject to long-term radiation and changes in temperature. The laboratories for this work were mainly grouped to the north-west. They included the Plutonium Fuel Laboratory A52, later known as the Alpha Materials Laboratory, a two-storey building constructed between 1961 and 1962 and commissioned in 1963 (Fig. 4.40). Specialised features in the building included 35 glove boxes, an automatic press, a continuous sintering furnace for uranium and plutonium, and a production line for sheathing these materials (Fig 4.41).[100]

Also in this area was the active handling building A59, constructed between 1962 and 1964, primarily to undertake post-irradiation studies on fuel assemblies and other activated components from the DRAGON and SGHWR reactors (*see* pp. 138, 140) (Fig. 4.42). During the 1970s, work expanded to include the examination of fuel from the CEGB's AGR fleet, and five caves were allocated to this work. This continued until 1992 when its facilities were used for some decommissioning tasks before it was finally closed in 1999.

The design of the building closely followed that of the Radio-Chemical Laboratory at Harwell. It contained two suites of heavily shielded caves with 1.52m-thick concrete walls and a shared rear cave access area (Fig. 4.43). The north line originally comprised five caves, later increased to seven, and the south line was made up of three. Operations in each cave were conducted by a pair of master–slave manipulators observed through zinc-bromide-solution-filled windows. For the handling and examination of more highly contaminated items the building was also equipped with a pressurised suit area (PSA).[101]

A pair of short railway sidings, just to the north of the Main Store A60 and garage, connected the site to the national rail network and were used for sending some types of spent fuel to Sellafield for reprocessing. A reliable water supply was another important factor in choosing a site. This was provided by two boreholes, and to the south-west a 1 million-gallon (4,546,090-litre) covered reservoir, 38.4m by 29.26m, was dug into Blacknoll Hill.[102] Liquid effluent was treated on site and held in storage tanks until it was released through two parallel steel pipes running nearly 10km to the sea at Arish Mell. These were formed of pipes within pipes, the inner pipe carrying slightly radioactive effluent for discharge 3.7km offshore and the other issuing effluent 15m below the spring tide low-water level.

Off site, the former Admiralty Underwater Weapons Launching Establishment, West Howe, Bournemouth, was acquired in July 1959 for £65,000, requiring a further £210,000 to equip it with specialised engineering machinery. Work was carried out under contract by the

Fig. 4.40
AEE Winfrith, Dorset, A52 Plutonium Fuel Laboratory, later known as the Alpha Materials Laboratory. [Courtesy of the Nuclear Decommissioning Authority]

Fig. 4.41
AEE Winfrith, Dorset, A52 Plutonium Fuel Laboratory glove boxes.
[Courtesy of the Nuclear Decommissioning Authority]

Fig. 4.42
AEE Winfrith, Dorset, A59, the active handling building.
[Courtesy of the Nuclear Decommissioning Authority]

Fig. 4.43
AEE Winfrith, Dorset, A59 Plutonium Fuel Laboratory, showing the manipulators used to handle material within the heavily shielded caves.
[Courtesy of the Nuclear Decommissioning Authority]

Hawker Siddeley Nuclear Power Company and from February 1961 by Vickers-Armstrong (Engineers) Ltd. This arrangement only lasted a few years before the work was taken in-house by UKAEA, and the West Howe works was sold in 1967.[103]

Research reactors and assemblies

In most instances, before work began on larger prototypes, experimental investigations were required in smaller research reactors to design fuel elements, core structure and the effects of radiation on reactor components. By the late 1950s, to improve the efficiency of reactor design, UKAEA was pursuing four possible options: advanced gas-cooled reactors (AGRs), fast-breeder reactors (FBRs), high-temperature gas-cooled reactors (HTGCRs), and steam-generating heavy-water reactors (SGHWRs). The variety of private companies involved in their procurement also illustrates the developing supply chain needed to support this new industry.

Work at Winfrith contributed to all four research strands, and smaller reactors and subcritical assemblies were installed in the central complex of buildings (Table 4.3). Subcritical assemblies occur where a nuclear fuel is releasing insufficient neutrons to maintain a self-sustaining reaction, or criticality. In some experimental situations, external neutron sources might be introduced.

One of its earliest research priorities was to gain a deeper understanding of HTGCRs. To support this research, ZENITH I, a zero-energy high-temperature gas-cooled reactor was designed by the Authority and manufactured by the General Electric Co. Ltd through their Atomic Energy Division at Erith, for installation in Zero Energy Hall No. 2 (B32) to operate under conditions similar to those that might be experienced in a high-temperature reactor.

Its core was about 1.22m in diameter and was made up of 235 fuel elements that comprised graphite sleeves holding pellets of thorium oxide, uranium oxide and graphite. The ends of the fuel elements were graphite, and with a 0.91m-thick graphite annulus surrounding the core formed a reflector to return some of the escaping neutrons to the core. Heated nitrogen gas was then passed through the core at 800 degrees Celsius before being mixed with cooler gas to keep the reflector at a maximum

Table 4.3 AEE Winfrith reactors and subcritical assemblies.

	Name	*Type*	*Fuel*	*Moderator*	*Coolant*	*Operational*	*Shut down*	*Typical maximum power (MWe)*	*Notes*
1	SGHWR	Steam-generating heavy-water reactor	Enriched uranium	Heavy water	Light water	1967	1990	100MW	
2	DRAGON	High-Temperature Gas-Cooled Experimental Reactor	Uranium-235 and Thorium, plutonium-239	Graphite	Helium	1964	1976	N/A	
3	ZENITH I	Zero-Energy Nitrogen-Heated Thermal Reactor	Enriched uranium and thorium, later plutonium	Graphite	Nitrogen	1959	1964		
4	ZENITH II	Zero-Energy Thermal Reactor	Variable	Graphite		1972			
5	NERO	Na (sodium) experimental reactor, '0' zero energy	Various, slightly enriched uranium	Graphite		1960	?1963		Ex-Harwell
6	JUNO		Various	Light or heavy water		1964			Reused NERO's foundations
7	NESTOR	Neutron-Source Thermal Reactor	Enriched uranium	Light water	Light water	1961	1995		
8	DIMPLE	Deuterium-Moderated Pile Low Energy	Various	Heavy water	None	1962	1995	100W	Thermal reactor studies, ex-Harwell
9	ZEBRA	Zero-Energy Breeder Reactor Assembly	Enriched uranium, plutonium	None	None	1962	1983		
10	HECTOR	Heated Experimental Carbon Thermal Oscillator Reactor	Various	Graphite	Carbon dioxide	1962			
11	HELEN	Subcritical	Various	Light water	None				
12	SCORPIO I	Subcritical Carbon-moderated Reactor assembly for Plutonium Investigations	Uranium-235 and plutonium	Graphite		?1963			
13	SCORPIO II	Subcritical Carbon-moderated Reactor assembly for Plutonium Investigations	Uranium-235 and plutonium	Graphite		?1963			

temperature of 400 degrees Celsius.[104] ZENITH II was also a zero-energy reactor and shared its control room with ZENITH I.[105]

Another early investigation was into the long-term behaviour of fuel rods in reactors leading to the burn-up of U-235 and the increase of plutonium. To address this question, two subcritical graphite-moderated reactor assemblies, known as SCORPIO I and SCORPIO II, were designed. They used special fuels manufactured by the Metallurgy Division at Harwell comprising depleted uranium, U-235 and plutonium, designed to mimic a partly used uranium fuel element from a reactor. Both were contained within steel shells, and when loaded, each assembly could be heated up to 400 degrees Celsius to measure reactivity at different temperatures. The reactors were initially housed in building B40 before being relocated to B20.[106] Another subcritical assembly, HELEN, was installed to study the reactor physics of light water, or other organic liquids, when used as moderators in combination with fuel using enriched uranium or plutonium. In late 1959 an order was placed with Hawker Siddeley Nuclear Power Co. for NESTOR (Neutron Source Thermal Reactor), a modified JASON design (*see* Box 4.2, p. 114), to act as a neutron source for subcritical assemblies of fuels and moderators. NESTOR was also located in B20 and went critical in March 1961; it operated until 1995.[107] The DIMPLE (Deuterium Moderated Pile of Low Energy) had been operating at Harwell since 1954 and was moved to B20 in 1961. It became critical in June 1962 and operated until 1995. Its applications included studies of light water, heavy water (*see* Box 2.1, p. 14) and other organic liquid moderators. It was also used to trial core structures for the steam-generating heavy-water reactor (SGHWR) (*see* p. 140).

In other experiments, the oscillator technique was used to measure power output as samples were moved in and out of a reactor. Initially, a zero-energy reactor NERO (New Experimental Reactor) was moved from Harwell to undertake this work, operating at room temperature in Reactor Hall B31. When NERO was shut down, most of its facilities were reused by a water-moderated reactor, JUNO, which went critical in April 1964.[108] To measure the effects on fuels, moderators and reactor materials over a wider temperature range, and neutron energy distribution, a new reactor, HECTOR (Heated Experimental Carbon Thermal Oscillator Reactor), a zero-energy reactor, was ordered from Fairey Engineering Ltd and H. M. Hobson Ltd. It was installed in Reactor Hall B32 and went critical in March 1962.[109]

Reactor groups

Widely spaced to the west and south of the administrative buildings and workshops were three distinct fenced reactor groups: DRAGON, ZEBRA and the SGHWR.

DRAGON

With so many projects underway, the UKAEA looked for ways of sharing research costs. Cockcroft, through his international contacts, was able to assemble a consortium of 12 European countries through the Organisation for European Economic Co-operation (OEEC) to build an experimental high-temperature reactor.[110] Managed through the European Nuclear Energy Agency (ENEA), Winfrith was chosen as the site of the experimental reactor named DRAGON. The agreement to proceed was signed in Paris in March 1958 and work began on 1 April 1959, with the project designed to run for five years (Fig. 4.44).

The headquarters for the DRAGON project was at the extreme eastern end of the site, and to hasten construction the general specifications for the building were set by R. S. Brocklesby of UKAEA, and a 'package' contract was let to the Turriff Construction Corporation to undertake detailed design work and construction (Fig. 4.45). The structure was T-shaped in plan, three storeys tall, with a four-storey service wing; the main offices were formed from precast concrete using a waffle-slab construction designed by Stephen Revesz and executed by Modular Concrete Ltd.[111]

The DRAGON reactor compound lay at the opposite end of the site centred on a gasometer-like reinforced containment structure standing 26.21m tall, with a further 1.89m below ground (Fig. 4.46). The supporting buildings comprised a three-storey control block linked to the containment structure by a raised walkway with access through an airlock, a separate services block, an active storage block, and a cooler block with a stack using air-blast heat exchangers as there was no facility to use the heat generated by the core to produce electricity. To create a secure foundation for the reactor a circular hole 36.58m in diameter at its base and 12.19m deep was excavated and filled with reinforced concrete (Fig. 4.47). The main containment building was 33.53m in diameter and 33.68m tall above the

foundation slab. It was formed from two concentric concrete walls 4.57m apart, each about 0.61m thick, with five floor levels. Inside the inner wall, the reactor, its main support equipment and heat exchangers were contained within a welded steel containment vessel (Fig. 4.48). Where extra protection was needed around the reactor, barytes and shot concrete was used. The core comprised 37 fuel elements, which were usually formed of seven fuel-bearing rods. These were filled with coated fuel particles of enriched uranium and thorium, about 1mm in diameter, in a graphite matrix, which acted as a can and moderator.[112] The graphite was manufactured by

Fig. 4.44
AEE Winfrith, Dorset, the DRAGON compound in 2010, showing the circular containment vessel with the disconnected end of the walkway. To the right is the control block, to the left the services block and to its rear the cooler block with a vent stack.
[© Historic England Archive, 26652_002]

Fig. 4.45
AEE Winfrith, Dorset, DRAGON project headquarters. The offices were formed from precast components, while the four-storey services block was cast in situ to permit early installation of the plant.
[Courtesy of the Nuclear Decommissioning Authority]

Fig. 4.46
AEE Winfrith, Dorset, DRAGON reactor complex designed by UKAEA architects T. L. Viney and R. S. Brocklesby. [Courtesy of the Nuclear Decommissioning Authority]

AERE's Bracknell works.[113] The reactor was controlled by 24 boron carbide absorber rods. Work began on site in April 1960 and it achieved its first criticality in August 1964, an international first for this type of reactor. The reactor buildings were designed by T. L. Viney and R. S. Brocklesby of UKAEA's architects department at Risley.

With successive project extensions, DRAGON operated until 31 March 1976. Its shutdown marked the United Kingdom's withdrawal from high-temperature reactor technology, with future research concentrating on steam-generating heavy-water reactors and fast-breeder reactors.[114]

ZEBRA

To support the development of the Prototype Fast Reactor (PFR) at Dounreay, building on work in the ZEPHYR and ZEUS reactors at Harwell, ZEBRA (Zero-Energy Breeder Reactor Assembly) was constructed to the south of the huts of the Reactor School and commissioned on 19 December 1962 (Fig. 4.49). It was designed as a zero-energy reactor in which different shapes and compositions of core assemblies might be trialled, and was housed in a circular containment structure with a flattened hemispherical roof, D50, surrounded by supporting infrastructure and offices. It was also used to investigate the neutron physics of fuel assemblies containing uranium and plutonium. To study a variety of options for core designs, experimental cores were assembled from up to 400,000 plates of materials that might be found in a reactor, including natural uranium, enriched uranium, plutonium, sodium and steel. Each plate measured 50mm square and 3mm thick and these were packed into 3m-long steel section tubes. The plates were fabricated on site in the Plutonium Fuels Building A52. Later it was involved in international research, collaborating with a Japanese team between 1971 and 1973 to develop a fast reactor in Japan. Subsequently, as part of a European project to develop a second-generation fast-reactor station, ZEBRA was enlarged for the BIZET (Big ZEBRA Experimental Test).[115] The facility operated until 1982, since when its site has since been cleared.

Steam-generating heavy-water reactor (SGHWR)

From the late 1950s uncertainties remained about the future development potential of gas-cooled reactors, in particular regarding the effects of long-term irradiation on their graphite moderators and the challenges of fabricating large steel pressurised vessels. One system that might provide an alternative for the third-generation commercial reactor fleet was steam-generating heavy-water reactor (SGHWR) technology. Previously in the United Kingdom relatively little work had been

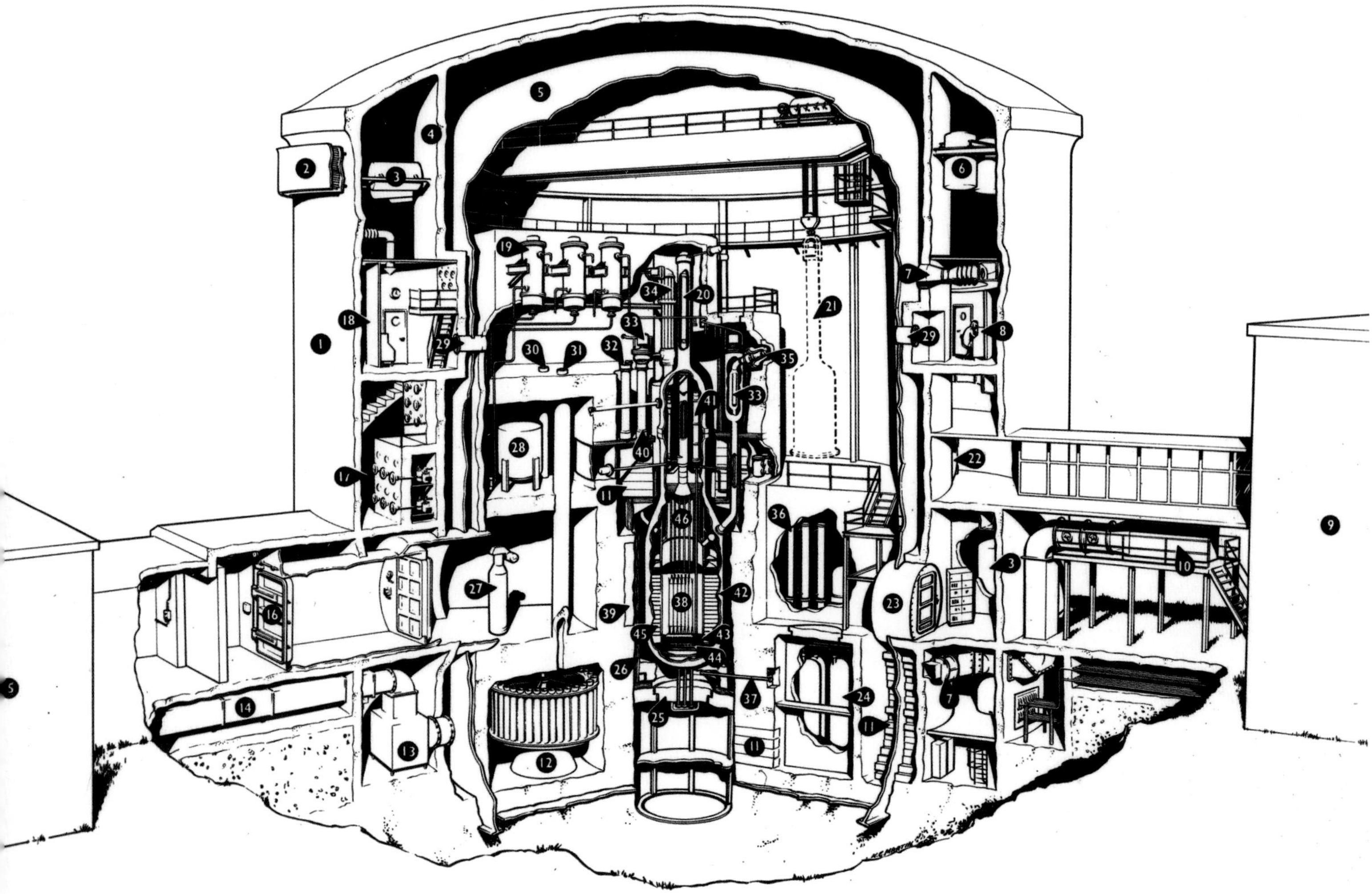

Fig. 4.47
AEE Winfrith, Dorset, DRAGON diagram.

Key
1 Outer concrete containment
2 Shut-down coolers
3 Water storage tanks
4 Inner concrete wall
5 Inner containment steel shell
6 Nitrogen storage tanks
7 Circulating ventilation plant
8 Crane control room
9 Control block
10 Main ventilation inlet and filters
11 Loose shield blocks
12 Fuel-element storage roundabout
13 Ventilation exhaust filters
14 Ventilation exhaust to stack
15 Active storage block
16 Vehicle airlock entrance
17 Secondary coolant control valves
18 Charge machine control room
19 Secondary heat exchangers
20 Charge machine
21 Charge machine in test area
22 Personnel airlock access
23 Emergency personnel airlock access
24 Helium dump tanks
25 Bottom biological shield
26 Iron water thermal shield
27 Transit flask
28 Secondary coolant dump tank
29 Observation window
30 Fuel-element charge chute
31 Fuel-element discharge chute
32 Can-washing tank
33 Primary heat exchanger
34 Load facility transfer flask
35 Primary circulation
36 Fission-product removal plant
37 In chambers
38 Core
39 Neutron shields
40 Canning induction furnace
41 Storage can on CM turntable
42 Main pressure vessel
43 Core bedplate
44 Fission product pipes
45 Reflector
46 Absorber rods

[Courtesy of the Nuclear Decommissioning Authority]

Fig. 4.48
AEE Winfrith, Dorset, DRAGON, interior of reactor hall. [Courtesy of the Nuclear Decommissioning Authority]

Fig. 4.49
AEE Winfrith, Dorset, ZEBRA. To the left is the reactor containment vessel and to its right supporting buildings. These buildings have been demolished. [Courtesy of the Nuclear Decommissioning Authority]

undertaken on heavy-water-moderated reactors for power generation, in part due to the cost of acquiring this material and the requirement for enriched fuels.[116] Concerns also remained about future uranium supplies. However, the ability of the SGHWR technology to use a variety of fuel types, including plutonium produced by the Magnox fleet, was an attractive prospect, as was the capability to more readily vary the output of this type of reactor to respond to changing load demands by the grid.

Approval for a prototype was given in February 1963 and, after countering initial objections by the county council that this reactor would change the nature of the establishment from research to industrial, planning permission was granted in March. Construction began in May, with building work completed in early 1967. The reactor achieved criticality in September and reached full power of 100MW of electricity in January 1968. This was despite an apparent case of industrial sabotage discovered in March 1964, when traces of mercury and other harmful materials were found in the reactor. An investigation was launched, and although it was determined they had been deliberately introduced, no perpetrators were identified.[117]

The SGHWR's compound was located at the extreme south-west corner of the site and was dominated by the largely windowless reactor and turbine hall clad in blue and white metal sheeting (Figs 4.50 and 4.51). On the northern side was a two-storey annexe housing the control room, changing room and electrical switch gear. To break up the large expanse of metal sheeting on the northern side, and wrapping around one corner, was a large 27.43m by 10.36m mural, an abstract design titled 'Fission', with swirling patterns in red, ochre, blue and black, symbolising the heat of the reactor. On the southern, rear, elevation of the building was a two-storey pumphouse. To the north of the reactor and turbine hall was the circular four-storey administrative block, connected to the main building by a first-floor-level walkway. To further reduce the visual impact of the structures, two banks of fan-assisted cooling towers were built to the south, the use of fans permitting lower structures to be erected. To the south-west was a switch-gear compound connecting the facility to the National Grid.

The reactor was fuelled by pellets of uranium dioxide slightly enriched with the U-235 isotope; each was 1.07cm in diameter and 1.22cm

Fig. 4.50
AEE Winfrith, Dorset, steam-generating heavy-water reactor shown during decommissioning works in 2010. To the top, a gravel mound marks the site of the cooling towers and to the right is the switchgear compound.
[© Historic England Archive, 26652_004]

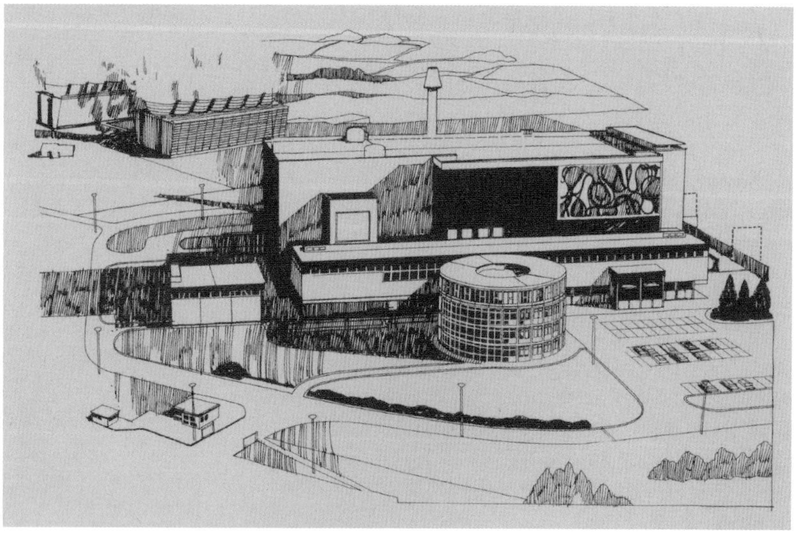

Fig. 4.51
AEE Winfrith, Dorset, 1980. Contemporary drawing of the steam-generating heavy-water reactor.
[Courtesy of the Nuclear Decommissioning Authority]

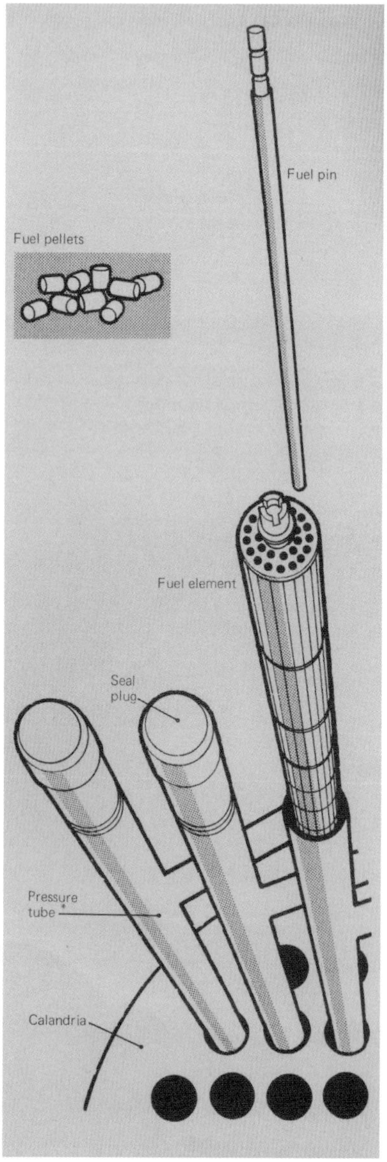

Fig. 4.52
AEE Winfrith, Dorset, steam-generating heavy-water reactor, diagram showing the position of a fuel element.
[Courtesy of the Nuclear Decommissioning Authority]

Fig. 4.53
AEE Winfrith, Dorset, steam-generating heavy-water reactor, the core coolant circuit.
[Courtesy of the Nuclear Decommissioning Authority]

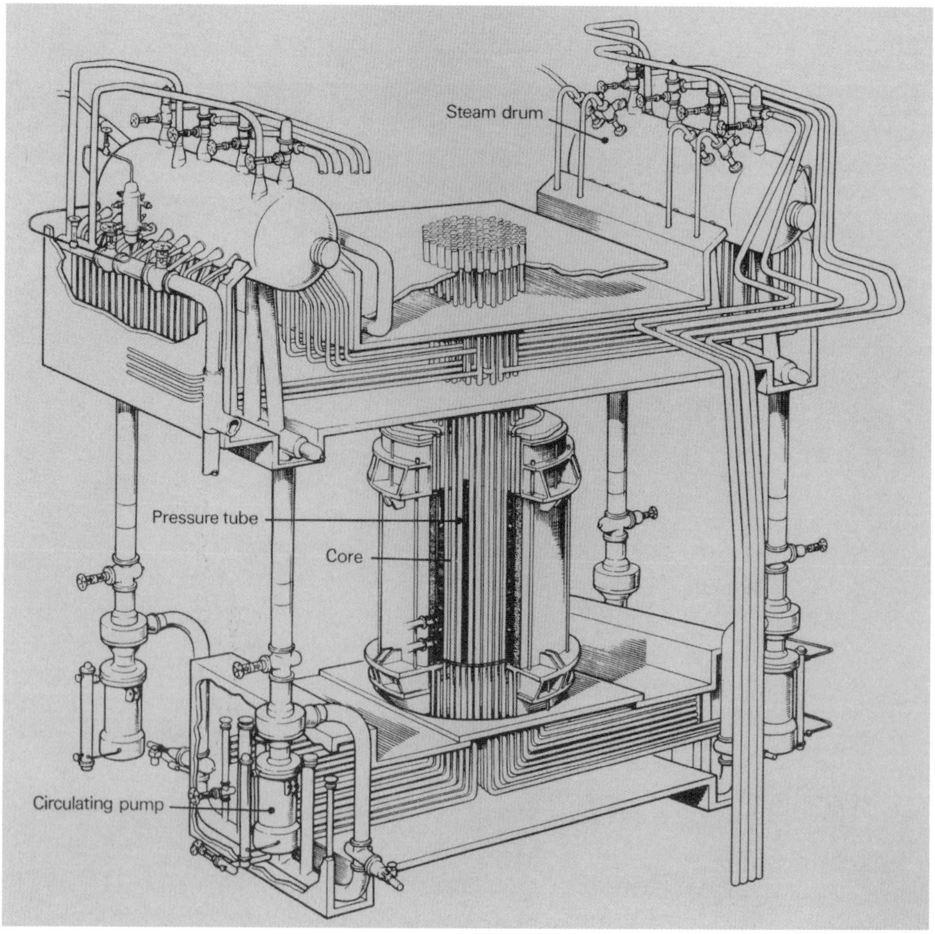

in length. They were loaded into 3.66m-long tubes, or pins, of zirconium alloy, a covering that absorbed fewer neutrons than stainless steel. Fifty-seven pins were assembled to create a fuel element; these were the same size as ones that might have been used in future commercial stations. The core comprised an aluminium tank, or calandria, 3.66m in diameter, filled with a moderator of heavy water. In this were 104 pressure tubes into which the fuel elements were inserted (Fig. 4.52). To control the reactor the level of the heavy water in the calandria could be varied. Shutdown was normally achieved by draining the heavy water into dump tanks beneath the reactor. In an emergency, to decrease reactivity, a liquid, referred to as a 'poison', which was capable of absorbing neutrons, might be injected into empty tubes before the heavy water was drained.

The primary cooling system comprised two independent and almost identical circuits (Fig. 4.53). It and the core were housed in the reinforced concrete containment vessel. In operation, water was taken from a steam drum by a circulating pump to an individual feeder pipe attached to each of the 104 pressure pipes. After partial evaporation, the water in each tube rose up the riser pipe to the steam drum where the water and steam were separated. The steam then passed through driers to the turbine. From the turbine condenser the cooled condensate then passed back to the steam drum. While shut down, the reactor was refuelled by a charging machine which removed the seal plug at the top of the pressure tubes before transferring the irradiated fuel elements to fuel transport flasks for reprocessing at Sellafield.[118]

The SGHWR was opened on 23 February 1968 by the Duke of Edinburgh in an atmosphere of great optimism (Fig. 4.54). It had been built to timetable and kept to its £16 million budget. There was also the prospect of export potential, with interest from Australia, New Zealand, Finland, Japan, and later Greece and

Fig. 4.54
AEE Winfrith, Dorset. To the rear is the steam-generating heavy-water reactor and to the right the AEI turbine generator. [Courtesy of the Nuclear Decommissioning Authority]

Italy. Such reactors were cheaper to build, as the direct steam cycle technology of the system did not require heat exchangers, and the use of heavy water as a moderator gave good neutron economy which lowered fuel costs. The design of the plant also permitted most of its equipment to be fabricated off-site by non-specialist manufacturers of steam and chemical plant.[119]

Although this early confidence was soon tarnished by technical issues resulting in shutdowns, in July 1974 it did not deter Eric Varley, Secretary of State for Energy in the Wilson Labour government, from choosing the SGHWR option for new stations proposed for Sizewell (Suffolk) and Torness, near Dunbar (East Lothian). Technical problems were not the only obstacle. By this time the CEGB was reporting a fall in the demand for electricity and saw no call for any new stations before the end of the decade. It also appeared less likely that any export orders would be forthcoming for SGHWRs to defray development costs. This caused consternation in the nuclear supply industry, which would be unable to survive without new orders. Independently, the CEGB had concluded that American Westinghouse pressurised-water reactors might best suit its requirements.[120] In spring 1978 Tony Benn, then Secretary of State for Energy, announced the end of funding for SGHWR research, although the Winfrith reactor continued to supply the grid until September 1990.[121] A promising technology had come to the end of the line, although it was emulated in an experimental plant at Fugen, Japan.

UKAEA, Culham

Early experimentation into fusion energy had begun at Harwell and Aldermaston in the mid-1950s, but as with reactor research, more space was required. In 1959 the UKAEA began to look for a new site to house its plasma physics and fusion research. Some of the site selection criteria included proximity to the Authority's existing sites and accessibility to the University of Oxford. The former Royal Naval Air Service airfield, HMS *Hornbeam*, at Culham (Oxfordshire) met these conditions, and on 29 January 1960 planning permission was granted for a research centre. The establishment was conceived, planned and constructed as a whole, its orientation and layout following that of the existing airfield's runways (Fig. 4.55). The buildings on

Fig. 4.55
UKAEA Culham, Oxfordshire. View from the south-west showing the 1960s establishment laid out on the orientation of the one of the airfield's runways.
[© Historic England Archive, OP04697]

the western part of the site survive essentially unaltered from the mid-1960s. These housed the United Kingdom's fusion research activities, including laboratories and offices for staff working on data analysis, theoretical studies, and other aspects of fusion research.[122]

Joint European Torus (JET)

In 1973, discussions began between the countries of the European Atomic Energy Community (Euratom) about the construction of a tokamak device (Russian for toroidal magnetic chamber) which would be larger than the United States' torus at Princeton and inform the construction of a prototype commercial fusion reactor. After difficult discussions, by late 1977 it had been agreed that the Joint European Torus (JET) would be housed at Culham with some services provided by UKAEA. The main construction contract was let to Tarmac National Construction, with advice on the electrical plant provided by the consultancy firm Merz and McLellan. The design of the structures was led by the architectural practice of Gordon Graham and Partners, with a track record of work in the energy sector. Reflecting its European origins, components were tendered for internationally and their diverse origins are evident throughout the building. The foundation stone was laid in May 1979 by EEC Energy Commissioner Guido Brunner, followed by a staggered build programme to ensure the main administrative and control buildings and the Torus Assembly Hall were complete by late 1980. The latter permitted the construction of the torus before its hall was complete.

Fig. 4.56 UKAEA Culham, Oxfordshire, Joint European Torus (JET) facility, September 2010. The tall building in the centre houses the torus, with the long rectangular raised sections on its roof marking the position of the shield walls, and to their rear is the roof laboratory above the tokamak. To the top left are the administration and control buildings, and to the right are the generator hall and power supplies building. [© Historic England Archive, 26744_018]

Fig. 4.57
UKAEA Culham, Oxfordshire, Joint European Torus (JET), southern elevation of the torus and assembly halls. The architectural design was led by Gordon Graham and Partners. On this side, the original blue panels, visible to the right, were later replaced with white panels to reduce thermal gain.
[© Historic England Archive, DP437669]

Fig. 4.58
UKAEA Culham, Oxfordshire, JET. The tokamak surrounded by diagnostic equipment.
[© Historic England Archive, DP437666]

The JET facilities nearly doubled the size of the existing UKAEA site and are dominated by the Torus Hall and Assembly Hall, with a combined footprint of 140m by 80m (Figs 4.56 and 4.57). The Torus Hall is in the western portion of the building and internally measures 35m square with a clear 23.5m height to the ceiling. To provide biological shielding against high-energy neutrons during deuterium/tritium experiments, the tokamak, its hall, access cell and adjacent hot cell are surrounded by reinforced concrete. The main hall walls are 2.5m thick, with an additional 0.3m of internal boron blockwork, the roof is 2.25m thick, and the floor is 1m thick, covering a basement holding ancillary equipment. To assist future dismantling, boronated screed was incorporated into the ceiling and floor to reduce the potential contamination of the main concrete wall. Penetrations through the walls and roof are provided to position diagnostic devices and for the introduction of gases into the tokamak.

The main entry from the Assembly Hall to the Torus Hall is through the Access Cell. This is protected by two massive sliding 400-tonne steel and concrete doors; a further shielding door also provides entry to the Hot Cell. To permit components to be transferred by the travelling crane from the Assembly Hall to the Torus Hall, the upper sections of the wall, above the shielding doors, comprise movable beams weighing 900 tonnes, which may be raised by hydraulic jacks into housings above the roof level.

At the centre of the Torus Hall is the tokamak, massive D-shaped magnets, 4m high and 2.5m across, which were used to create a very complex 3D magnetic containment field (Figs 4.58 and 4.59). Its function was to confine and control a plasma stream of fusion fuels, isotopes of hydrogen, deuterium and tritium, held under vacuum and heated by an electrical current to around 1 million degrees Celsius.[123] The aim was to cause the fusing of atoms, during which

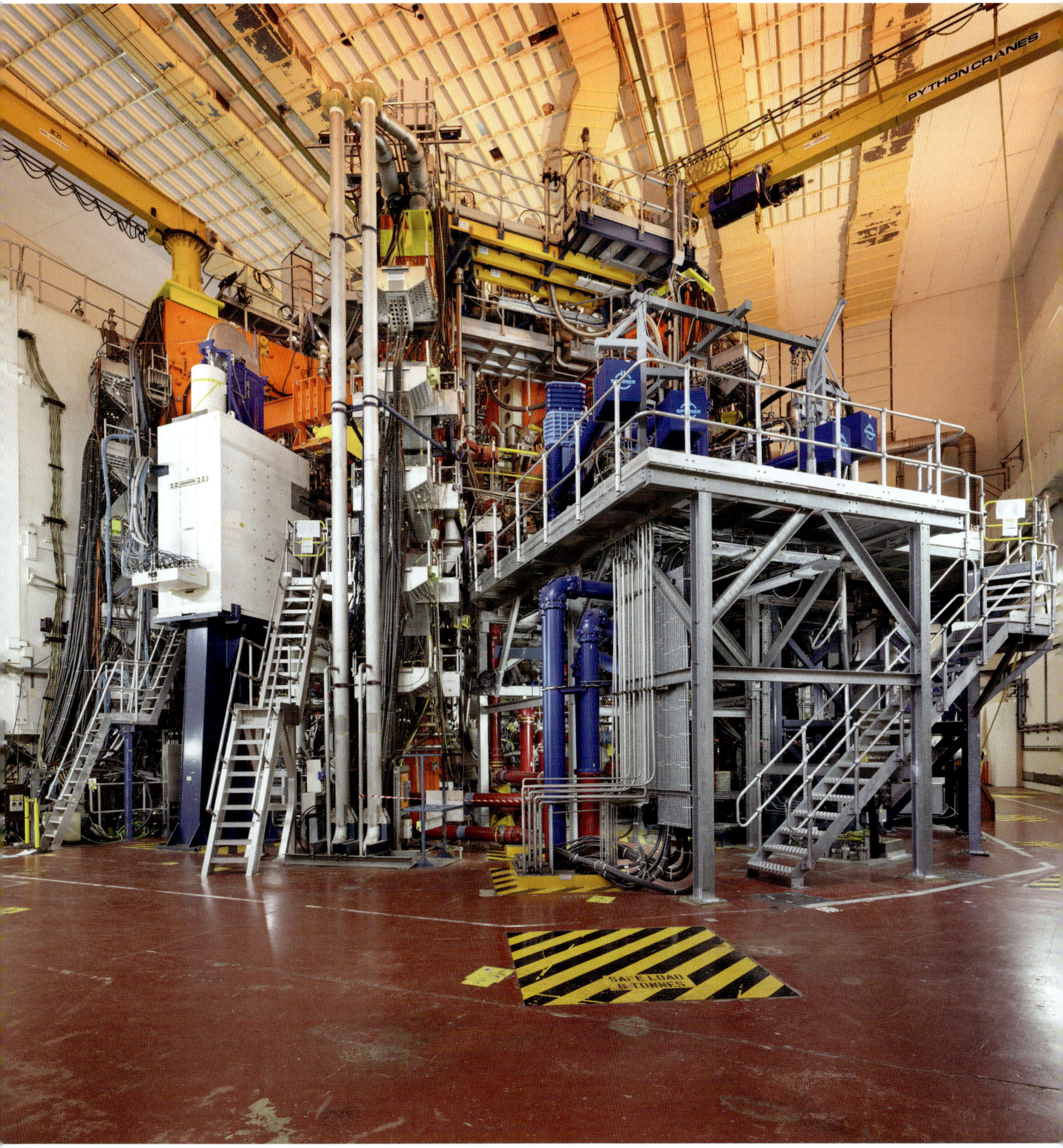
PYTHON CRANES
SAFE LOAD
6 TONNES

procedure more energy would be created than was put into the process. Encircling the tokamak was a bewildering array of rigs holding control and diagnostic equipment.

The Torus Hall in turn is surrounded by wings housing heating, power supplies, support facilities, and diagnostic equipment. To permit flexibility during construction and any later modifications, these are steel framed and clad externally in pressed steel sheeting, clear glazing and originally large patent-blue glass panels. To the east is the Assembly Hall, with a 35m span by 65m standing 23.5m tall; set in its roof is a 150-tonne travelling crane in addition to two smaller 20-tonne cranes. As its name suggests, the tokamak's components and supporting diagnostic equipment were brought together in this room prior to installation. It also houses a full-size model tokamak, with an unused segment, an octant, which is used as a training and development aid (Fig. 4.60).

To the south of the main building is the Control Building, measuring 50m by 35m, which is typical of buildings housing computer equipment, the corridor and many of the rooms having raised floors with cabling beneath. This building houses the main control room and data storage computers. To the north of the main building is the Generator Hall and the covered Static Electrical Equipment Buildings, and to their north the 400kV switchgear compound connected directly to the National Grid. To the north-east is a bank of forced-draught cooling towers.

When it became operational by June 1983, it was the largest fusion experiment in the world. In 1991 it was the first facility to produce controlled fusion power, and in late 2023, just prior to entering decommissioning, its final deuterium–tritium experiments consistently produced high fusion power for 5 seconds, setting a world record.[124]

In 1999, the Mega Ampere Spherical Tokamak Experiment (MAST) began operation at Culham, following on from the Small Tight Aspect Ratio Tokamak (START) experiment (1991–8). In the next development phase, a Spherical Tokamak for Energy Production

Fig. 4.59
UKAEA Culham, Oxfordshire. A model of the tokamak housed in the foyer of the JET administrative building. The figures at the base of the model indicate the size of the facility.
[© Historic England Archive, DP438092]

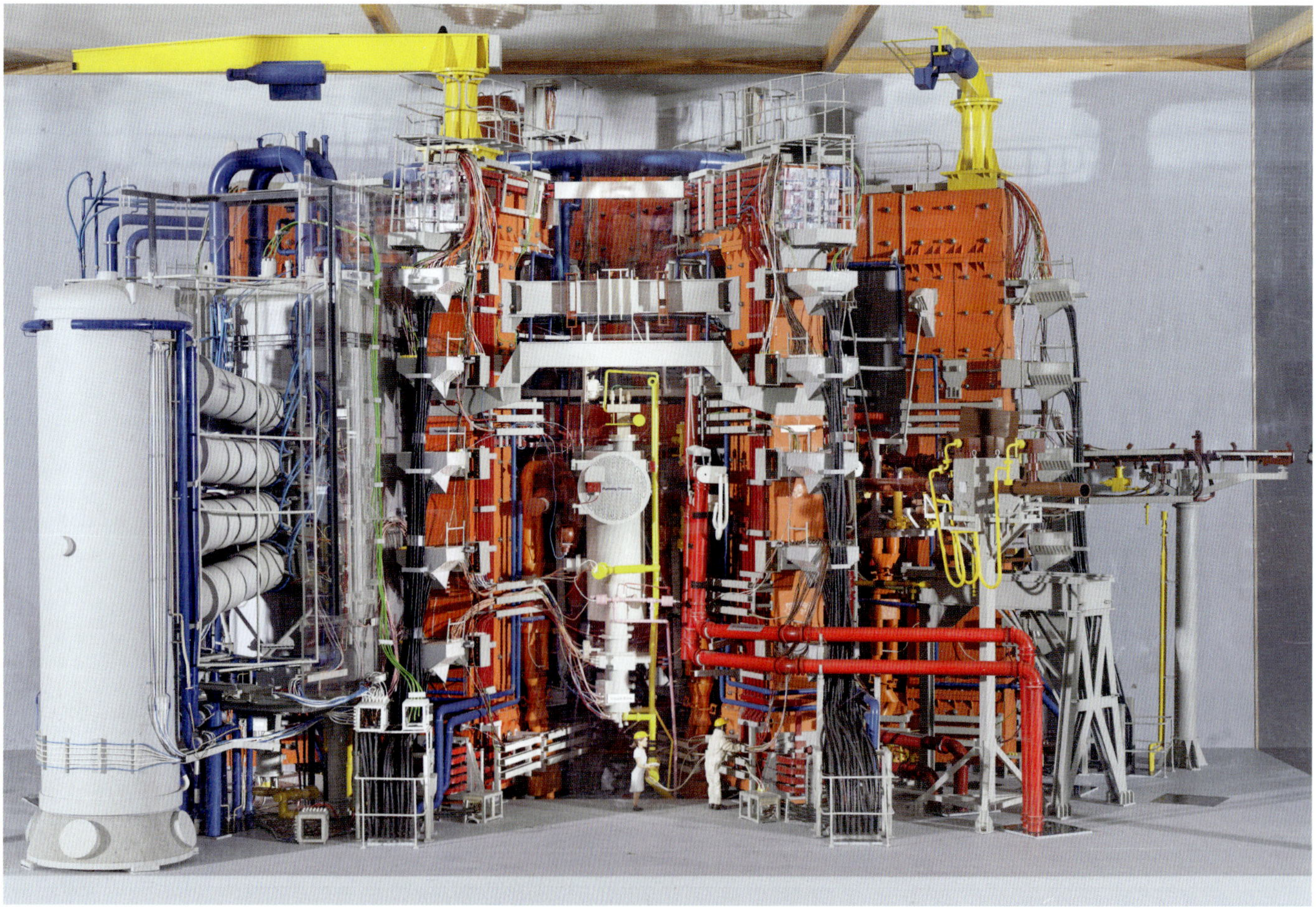

(STEP) plant will be built on the site of the coal-fired power station at West Burton, Nottinghamshire.[125]

Conclusion

Over the course of a few decades, investigations into nuclear physics had moved from small teams of scientists in outdated university laboratories to large, well-equipped research establishments. The Second World War had provided a successful model of how large, organised, government-funded research facilities employing hundreds of people had the ability to solve complex scientific problems and quickly produce practical applications; radar and the jet engine are two of the most obvious examples. This was the immediate experience of the politicians, managers and many of the personnel of the British scientific civil service who went on to lead the post-war nuclear research projects. It was a tremendously exciting time for scientists and engineers involved in this new field. But it was also a bewildering period, with seemingly endless potential combinations of reactor fuels, coolants and moderators that might create a successful reactor. The British research establishments were world class; with the exception of the American Idaho Testing Station, the greatest variety of research reactors in the world were at Dounreay, Harwell and Winfrith.[126] Beyond the government establishments, the capabilities of the country's engineering and electrical industries to manufacture high-precision plant was fundamental to this endeavour. In the next chapters we will discover how theoretical and applied research was applied to the creation of the civil nuclear power industry.

Fig. 4.60
UKAEA Culham, Oxfordshire, JET Assembly Hall, full-size training model of the tokamak incorporating an unused octant. This was used to simulate and perfect engineering procedures and operations before they were applied on the live system.
[© Historic England Archive, DP438105]

5

The first nuclear power programme

Introduction

On Tuesday 15 February 1955, the Minister for Fuel and Power, Geoffrey Lloyd, stood up in the House of Commons to present the White Paper *A Programme for Nuclear Power*, the world's first civil nuclear programme. This foresaw the construction of 12 nuclear power stations by 1965, generating 1,500–2,000MW. Not only was nuclear power envisaged as the 'energy of the future', but it was also seen as a means of filling a potential shortfall in coal supplies caused by the difficulty of recruiting miners. It was also anticipated that it would put the country at the forefront of civil nuclear technology, with the prospect of a lucrative export trade, by combining small amounts of imported raw materials with skill, ingenuity and inventiveness. Oil was becoming an increasingly important energy source, but, as the later Suez Crisis in autumn 1956 highlighted, it was based on a potentially vulnerable supply chain. One response in the following year was to treble the target for electricity generated from nuclear energy. However, by the end of the decade, an easing of fears about oil supplies and a more optimistic outlook for coal power generation led to a 'deferment of acceleration' for the nuclear programme.[1] This chapter will examine the development of Magnox reactor technology, other key station components, and how these influenced the construction of the stations, their architecture and ultimately their landscape setting. The Magnox programme took its name from the magnesium non-oxidisable alloy cans that encapsulated the uranium fuel rods (Fig. 5.1).

Fig. 5.1
Magnox fuel element, inert training aid, showing fins and helical grooves to aid heat exchange, on display at Sywell Aviation Museum. [© Historic England Archive, DP436974]

Today, the Magnox stations are closed and are being dismantled and demolished. The survival of components will vary between sites depending on where they lie in the decommissioning programme, but at all locations, lowered and reclad reactor buildings survive, housing the most heavily irradiated parts.

The Magnox programme

By the early 1950s, it was evident that the British government's weapons programme would require more plutonium. The Windscale Piles were running less efficiently than expected, partly due to demands to produce isotopes and other irradiated materials.[2] They were a new technology, and until their operational data was analysed, their lifespan was unknown, and with it their continuing ability to supply plutonium. One option to increase plutonium production might be to resume work on the abandoned third air-cooled Windscale pile. In considering reactor designs, the use of natural uranium fuel was the only fixed criterion. Constraints on the possible use of enriched fuels included the demand for plutonium by the weapons programme and the production capacity of the uranium enrichment plant at Capenhurst (Cheshire).

In the meantime, preliminary work by a handful of staff members at Risley (Warrington Unitary Authority) and Harwell (Oxfordshire) had continued to investigate gas-cooled reactors. In September 1950 a conference at Harwell had indicated that gas-cooled reactors might produce electricity as cheaply as traditional coal-fired stations, and that plutonium produced as a by-product during power production might later be used to power fast-breeder reactors.[3] In the following year, work on gas-cooled reactors was expanded beyond AERE to include the Ministry of Works, the Central Electricity Authority and the boiler makers Babcock and Wilcox. After several alternatives had been considered and engineering tests undertaken, it was decided to proceed with a reactor design comprising a graphite moderator cooled by carbon dioxide gas.[4]

This became one of the distinctive technological features of the British nuclear industry. As a coolant, carbon dioxide had many desirable

Fig. 5.2
Calder Hall, Cumbria, reactor group.

Key
1 Reactor
2 Turbine hall
3 Cooling tower
4 Administration
5 Electrical switchgear compound

[Contains OS mapping. © Crown copyright (and database rights) 2025 OS AC0000815036]

properties including inertness, abundance, relative cheapness, good heat transfer properties, and it is a relatively poor neutron absorber.[5] The idea of using gas cooling had first been proposed in the late 1940s for the design of the Windscale Piles. But, due to the necessity of producing plutonium as quickly as possible, the complex engineering challenges of designing a containment vessel and gas circulation system were thought to introduce too much delay, and a simpler air-cooled system was adopted (*see* p. 30).

Calder Hall, Cumbria

By 1952, R. V. Moore, AERE Harwell, had produced an initial design for a gas-cooled reactor known as PIPPA (Pressurised Pile Producing Power and Plutonium). Hinton and Cockcroft concurred with its potential, and Hinton proposed the idea to the Atomic Energy Council, chaired by the paymaster-general, Lord Cherwell, that this technology should be adopted, and that the reactor should be built adjacent to Windscale, at Calder Hall, named after an adjacent farm, now lost. It received government approval in early 1953, and soon after the design work begun it was agreed to build two reactors.[6]

Construction of the first pair of reactors at Calder Hall began in August 1953 (Figs 5.2–5.6).[7] They were housed in individual buildings but shared a turbine hall and two cooling towers. At the heart of each reactor was a 5.1cm thick cylindrical steel containment vessel with domed ends, measuring 21.3m high and 10.7m in diameter. This held the 1,219.26-tonne cylindrical graphite core, supported within the vessel on a steel framework known as the diagrid. The core was 8.2m high and 9.4m in diameter and comprised about 58,000 graphite blocks, each of which measured 0.2m square and 0.6m high with a 0.1m diameter hole through their centres. Once assembled, the core was pierced by about 1,800 holes, of which 1,696 were fuel element channels, each holding six fuel elements. At the centre of each group of 16 channels was an additional hole for a neutron-absorbing boron control rod, used to regulate output from the reactor and to act as a shutdown rod. Other channels held instrumentation and material samples that were periodically withdrawn to analyse the effects of irradiation.

Piercing the top of the dome were openings for 112 charge tubes, each of which served 16

Fig. 5.3
Calder Hall, Cumbria, May 1962, showing the four reactors and associated cooling towers. To the left is the River Calder.
[Courtesy of the Nuclear Decommissioning Authority, AJ3885]

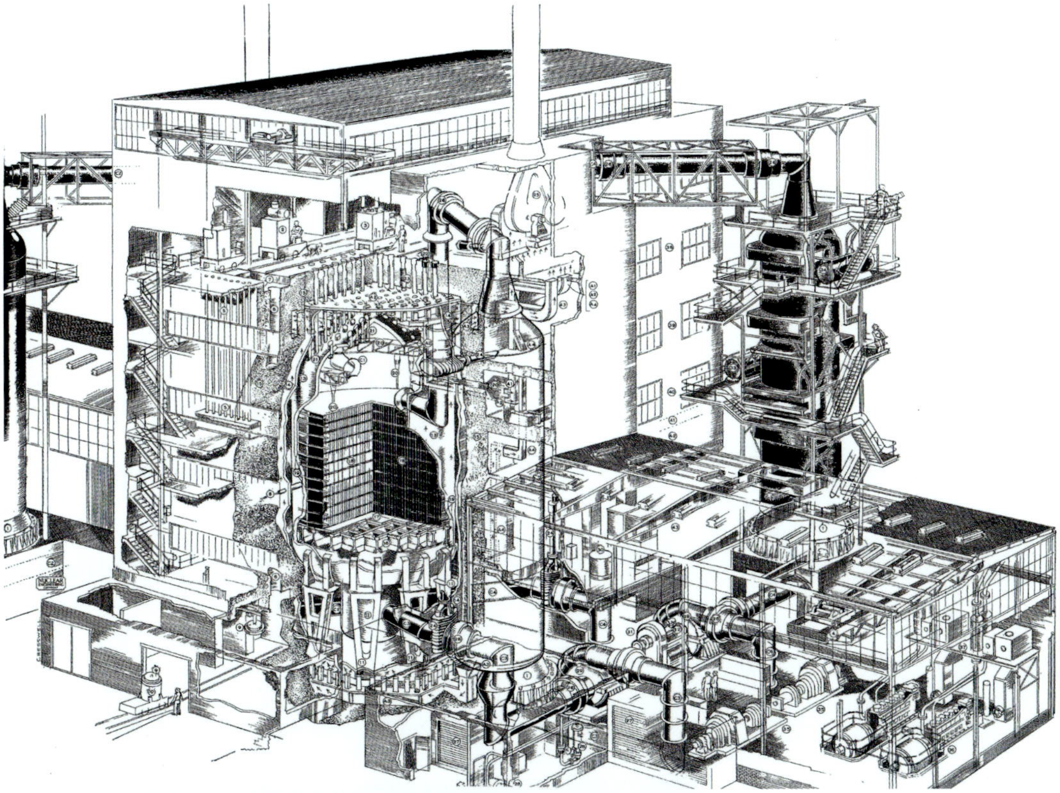

Fig. 5.4
Calder Hall, Cumbria. Diagram showing the main features of the reactor with the tall central reactor building surrounded by heat exchangers. [Courtesy of the Nuclear Decommissioning Authority]

Fig. 5.5
Calder Hall, Cumbria. Heat exchanger to left and reactor building to right. [Historic England Archive, OP03029]

Fig. 5.6
Calder Hall, Cumbria, control desk. Note the emergency red shut-down button and monitoring diagrams for the reactors and heat exchangers. [Courtesy of the Nuclear Decommissioning Authority, P-90-41892(3c) (30 Jul 1990)]

channels in the graphite core. These passed through the roof of the biological shield, which also formed the floor of the charge face. Above this was a travelling crane, with 7.6-tonne and 1.01-tonne hoists used for maintenance work and moving the charge and discharge machines. On the charge floor two rail-mounted charging machines, No. 1 painted yellow and No. 2 painted pink, held new fuel elements and a winch-mounted grab for lowering the elements into the fuel channels. Similar, although more heavily shielded, machines weighing about 71.2 tonnes were used for discharging spent fuel elements. To recharge, the reactor was shut down and a charge tube prepared by withdrawing the shield plug and inserting a guide chute to bridge the gap between the underside of the charge floor and the pressure vessel. After removal, spent fuel elements were then lowered down a chute to a water-filled, lead-lined steel container. At Calder Hall the elements were then taken directly to the reprocessing plant (*see* p. 174), obviating the need for a cooling pond.[8]

The development of Magnox reactor technology

Calder Hall provided the template for all the Magnox reactors that followed, although each new development brought refinements and sometimes more significant innovations based on a growing body of operational and scientific knowledge. Although this approach produced an impressive increase in power output (Tables 5.1–5.3), the mix of consortia employed to build the stations resulted in little standardisation of components – a consequence of the policy to spread expertise across an emerging British nuclear industry. The rescheduling of the programme in 1960, coupled with more powerful stations, reduced the need for as many stations and resulted in the reduction of construction consortia from five to three (Tables 5.2 and 5.3).

In June 1955, to increase plutonium production, an additional pair of reactors was approved for construction to the north of the original Calder Hall station, and a second reactor site was chosen at the former airfield at Chapelcross (Dumfries and Galloway) (Fig. 5.7).[9] To hasten the construction of this new station, its design closely followed that of Calder Hall, and many of the same components and contractors were used.[10] Here the line of the runway provided the main axis of the station; the open site also facilitated a more efficient layout during construction and operation. Minor alterations to the pressure vessel included changes to the design of the sampling pipes and improvements in the operation of the control rods. The 1957 Windscale fire had raised serious questions about the use of graphite as a moderator and ways in which the Wigner effect (*see* p. 34) might be controlled. At Chapelcross, graphite sleeves were placed

Fig. 5.7
Chapelcross, Dumfries and Galloway. Postcard of the station showing the four reactor buildings and cooling towers. Here the private practice of L. J. Couves and Partners was responsible for its design. [Author's collection]

Table 5.1 Magnox stations technical summary. Power outputs are indicative and varied through time due to modifications to turbines or reactors being run at less than full power.

Station	*Operational*	*Design rating*	*Reactors*	*Turbines*
Calder Hall	1956–2003	192MW	4	C. A. Parsons & Co. Ltd 4 × 27MW
Chapelcross	1959–2004	140MW	4	C. A. Parsons & Co. Ltd 8 × 23MW
Berkeley	1962–89	275MW	2	Associated Electrical Industries Ltd 4 × 83MW
Bradwell	1962–2002	300MW	2	C. A. Parsons & Co. Ltd 6 × 5MW
Hunterston	1964–89	360MW	2	General Electric Co. Ltd 6 × 60MW
Hinkley Point A	1965–2000	500MW	2	English Electric 3 × 93.5MW
Trawsfynydd	1965–91	500MW	2	Richardsons Westgarth & Co. 4 × 145MW
Dungeness A	1965–2006	550MW	2	C. A. Parsons & Co. Ltd 4 × 142.5MW
Sizewell A	1965–2006	580MW	2	English Electric 2 × 325MW
Oldbury-on-Severn	1967–2012	600MW	2	C. A. Parsons & Co. Ltd 2 × 313MW
Wylfa	1971–2015	1180MW	2	English Electric 4 × 335.1MW

Data source: https://assets.publishing.service.gov.uk/media/5a74eb21e5274a59fa715fc7/Table_of_past_and_present_UK_nuclear_reactors.pdf (acc. 17 January 2024).

Table 5.2 Atomic power consortia, 1957.

Consortia	GEC/Simon-Carves Atomic Energy Group	AEI-John Thompson Nuclear Energy Co. Ltd	Nuclear Power Plant Co. Ltd	English Electric Co. Ltd	Atomic Power Construction Ltd
Constituents	General Electric Co. Ltd, Fraser & Chalmers Ltd, Simon-Carves Ltd, The Motherwell Bridge & Engineering Co. Ltd, John Mowlem & Co. Ltd	Metropolitan-Vickers Electrical Co. Ltd, British-Thomson Houston Co. Ltd, John Thompson Ltd, Balfour Beatty & Co. Ltd, John Laing & Sons Ltd	C. A. Parsons & Co. Ltd, A. Reyolle & Co. Ltd, Head, Wrightson & Co. Ltd, Sir Robert McAlpine & Sons Ltd, Whessoe Ltd, Strachan & Henshaw Ltd, Alex Findlay & Co. Ltd, Clarke, Chapman & Co. Ltd	Babcock & Wilcox Ltd, Taylor Woodrow Ltd Group	Richardsons Westgarth & Co. Ltd, International Combustion (Holdings) Ltd, Crompton Parkinson Ltd, Trollope & Colls Ltd, Holland & Hannen & Cubitts Ltd

Source: Oswald, W. 1957 'Industry's Contribution to Nuclear Power', BEAMA Journal (February), 7.

around the fuel elements in three out of the four reactors. This created a small gap between it and the main core, which was filled with carbon dioxide. This allowed the temperature of the main core to be raised, in turn producing some self-annealing to overcome the Wigner effect.[11]

Construction of the first commercial stations at Berkeley (Gloucestershire) (Fig. 5.8) and Bradwell (Essex) commenced in January 1957, although progress was delayed while lessons from the 1957 fire were assessed and design modifications made. One of the factors that affected the efficiency of reactor operation was the pressure of gas passing through the containment vessel and its ability to resist this force. At Berkeley, although larger, the containment vessels were of a similar design to Calder Hall, comprising a cylinder with domed ends. Here, advances in welding technology allowed the use of 7.62cm thick steel plate, which in turn

Table 5.3 Main British nuclear power consortia, 1961.

Revised consortia	English Electric, Babcock & Wilcox, Taylor Woodrow Atomic Power Construction Co. Ltd	The Nuclear Power Group	United Power Co. Ltd
Notes		Was the AEI-John Thompson Nuclear Energy Co. Ltd and the Nuclear Power Plant Co. Ltd	A merger of Atomic Power Constructions Ltd and the GEC/Simon-Carves Atomic Energy Group
Constituents	Babcock & Wilcox Ltd, English Electric Co. Ltd, Taylor Woodrow Construction Ltd	Associated Electrical Industries Ltd, Clarks, Chapman & Co. Ltd, Alex Findlay & Co. Ltd, Head Wrightson & Co. Ltd, Sir Robert McAlpine & Sons Ltd, C. A. Parsons and Co. Ltd, A. Reyolle & Co. Ltd, Strachan & Henshaw Ltd, John Thompson Ltd, Whessoe Ltd	Fairey Co. Ltd, General Electric Co. Ltd, International Combustion (holdings) Ltd, Richardsons, Westgarth & Co. Ltd, Simon-Carves Ltd
Headquarters	Whetstone, Leicestershire	Radbroke Hall, Knutsford, Cheshire	28/30 Theobalds Road, London
Contracts	Hinkley Point	Bradwell	Hunterston
	Sizewell	Berkeley	Trawsfynydd
	Wylfa	Dungeness	Tōkai-Mura (Japan)
		Oldbury	
		Latina (Italy)	
Research laboratories	Whetstone, Leicestershire (EE Co.) Renfrew (Babcock)	Aldermaston Court (AEI) Newcastle upon Tyne (Parsons)	Heston, Middlesex (APC Ltd) Erith, Kent (GEC)

Source: *Electrical Review*, 17 November 1961, 778.

Fig. 5.8
Berkeley, Gloucestershire, August 1993. Two reactors surrounded by heat exchangers. To the top left are the Berkeley Nuclear Research Laboratories. [© Crown copyright. Historic England Archive, 298210 4845_25]

permitted the internal pressure to be raised. They were manufactured by the boilermakers John Thompson of Ettingham, Wolverhampton, and at 1,016 tonnes they were at the time the largest steel vessels constructed.[12] Additional safety features included 1,000 thermocouples to monitor temperatures across the core, and about 9,000 graphite samples, which could be periodically removed for examination.[13]

Another major advance in reactor operation at Berkeley was the ability to charge and discharge the reactor while it was on load (Fig. 5.9). This was a significant development for stations designed to deliver constant base-load power to the grid, and to meet early, but unfounded, concerns about the reliability of the fuel elements.[14] On the charge face were seven machines, three combined charge and discharge machines, two chute machines, a control rod and actuator servicing machine, and a visual observation and removal machine.[15] To refuel the reactor one of the concrete-filled standpipes would be removed and the chute machine used to position various chutes between the charge face and the reactor. The chute machine was then withdrawn, and the charge and discharge machine placed over the chute. Internally, its magazine had the capacity to hold enough new and spent fuel elements for a single channel.[16] Used elements were lowered down a chute and then moved into a cooling pond before being transported to Windscale for reprocessing.

At Bradwell (Figs 5.10–5.12), a spherical reactor vessel of 7.6cm thick steel plate was introduced, a form used in the remainder of the Magnox stations until the adoption of pre-stressed concrete vessels. This design allowed the reactor to work at higher inlet pressure and outlet pressures.[17] Here charging and discharging could also be carried out while the reactor was operating. A single machine was used for this operation, with another machine held on stand-by (Fig. 5.13). From the machine, spent fuel elements were transferred through a shielded tube to an underwater skip for transfer to a separate cooling pool, similar to the one illustrated here at Sizewell (Fig. 5.14) before they were transported for reprocessing.

At Hunterston (Ayrshire), the South of Scotland Electricity Board built the country's

Fig. 5.9
Berkeley, Gloucestershire, Reactor No. 1, charge face with, to the left, the chute machine, and, in the centre, the control rod and actuator servicing machine, with overhead travelling crane above.
[© Historic England Archive (John Laing Photographic Collection), JLP01_08_065023]

only Magnox station. It was an ambitious design comprising a pressure vessel with a double shell, the inner temperature resistant, and the outer non-brittle. Another unique feature of this station was its fuel-handling procedures. Charging and discharging were carried out in a room located beneath the reactor (Fig. 5.15). The advantages of this arrangement were that these operations were conducted at the coolest part of the reactor, and it separated the fuel handling from reactor maintenance activities. Three machines were provided so that one could be removed for servicing. Standpipes projected down to the roof of the charging chamber and were normally sealed by steel and concrete plugs. Beneath these, the combined charge and discharge machine was mounted on rails and a turntable that allowed it to be positioned beneath any of the fixed channel tubes. The machine was also used for changing the position of fuel elements to ensure the most efficient use of the uranium. It was so designed that it could operate while the reactor was on load and could remove and load a complete fuel channel of ten fuel elements.[18] To ensure the most efficient use of the fuel, this machine might also be used for axial inversion of the fuel elements. In this process they were removed from the fuel channels and reordered so that the more heavily irradiated central elements were placed at the ends of the channels, and the less irradiated elements in the centre.[19] Above the upper floor was a travelling crane used to position the reactor servicing machine over the standpipes. It might also be used for more specialised tasks, such as the recovery of fallen control rods and jammed fuel elements, the insertion of fuel elements with thermocouples or a monitoring camera, and the emergency removal of fuel elements.[20]

The station destined for Hinkley Point (Somerset) represented a substantial increase in power output to 500MW and was at the time the world's largest reactor. This growth was achieved by increasing the operating pressure and mass gas flow, by more efficient use of the space inside the pressure vessel by the core, and by using a smaller lattice pitch – the distance between the centres of neighbouring fuel rods.[21] Fuelling could also be carried out while the reactor was on load. This was achieved by a hole preparation machine, which prepared the link to the reactor for the fuelling machine, the latter being fitted with a pneumatic grab, winding gear and magazines for new and irradiated fuel. Both machines were manoeuvred into position by an overhead crane.[22] At Trawsfynydd (Gwynedd), advances in welding technology allowed the thickness of the spherical pressure vessel to be increased to 8.9cm, in turn permitting a further increase in internal gas pressure. Modifications in the design of the core also gave a more efficient ratio between core volume and pressure vessel volume.[23] Each reactor was served by two heavily shielded charge machines, which could be connected to any of the standpipes.

Fig. 5.10
Bradwell, Essex, August 2010. To the right are the reactor buildings and to their left the turbine hall, and to the bottom-left the electrical substation.
[© Crown copyright. Historic England Archive, 26722_053]

Fig. 5.12
Bradwell, Essex. Diagram showing principal buildings.

Key
1 Reactor building
2 Turbine house and control room
3 Cooling pond and pond water treatment
4 Circulating water pump house
5 Administration, welfare and canteen
6 Main stores and workshop
7 Visitor centre
8 Switchgear compound

[Contains OS mapping. © Crown copyright (and database rights) 2025 OS AC0000815036]

Fig. 5.11
Bradwell, Essex, June 1962. View of the reactor buildings from the electrical substation.
[Courtesy of the Nuclear Decommissioning Authority, AJ3900]

Fig. 5.13
Bradwell, Essex. View of a reactor floor with the 406.4-tonne, 15.85m-tall charge machine. Mounted on a gantry, it could be moved across the pile cap to a specified standpipe where it was able to load and remove fuel elements, absorber rods and control rods.
[CEGB, 1181]

Fig. 5.14
Sizewell, Suffolk. Cooling pond shared by both reactors. Fuel elements were stored in these ponds for many months before being transferred to Sellafield for reprocessing. [CEGB, 76082]

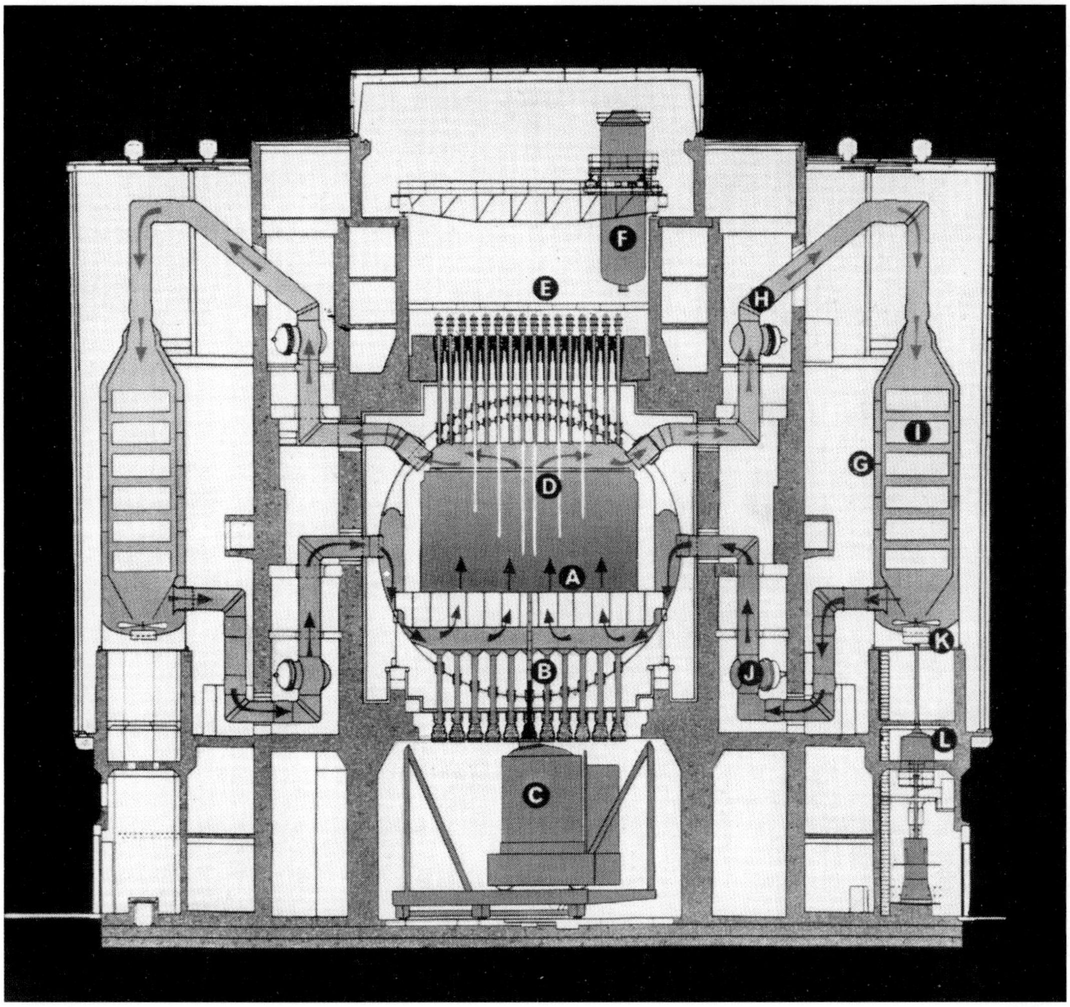

Fig. 5.15
Hunterston A, Ayrshire. At the centre, reactor building is a 21.34m-diameter spherical steel pressure vessel 7.6cm thick and weighing 1,828.8 tonnes. It contained the 28-sided, 15.39m-diameter, 8.5m-tall core pierced by 3,288 fuel channels, mounted on the diagrid support. Above the vessel, vertical standpipes pass through the upper shield, which contain control rods and provide access to the reactor for servicing. This was carried out by the heavily shielded servicing machine that was manoeuvred into place by the travelling crane. Unusually, the charge machine is located beneath the reactor where it uses another set of standpipes. Pipes exiting from the top of the reactor carried the heated carbon dioxide to the heat exchangers. After passing through the heat exchangers, or boiler, to repeat the cycle it was pumped back into the core by the gas circulators. From the heat exchangers steam passed through a pipe to the turbine house.

Key
A Reactor core
B Fuel loading
C Charge and discharge machine
D Control rods
E Control rod mechanisms
F Service machine
G Steam raising unit (8 per reactor)
H Heated carbon dioxide
I Steam-raising units
J Cooled carbon dioxide
K Carbon dioxide circulator
L Circulator motor

[Source: G. A. Mears n.d. Hunterston: A Nuclear Power Station *(South of Scotland Electricity Board), 7]*

The design of Dungeness A (Kent) (Fig. 5.16) by The Nuclear Power Group was informed by their experiences at Berkeley and Bradwell, and the British-designed Magnox station built at Latina, Italy. Further improvements in welding technology allowed the thickness of the base of the pressure vessel to be increased by up to 11cm. One method of reducing the build-up of Wigner energy was to increase the operating temperature of the core. This could be achieved by raising the gas inlet temperature, although this affected the thermal efficiency of the reactor by reducing the temperature rise as the gas flowed through the reactor. The build-up of Wigner energy might be countered by increasing the gas flow or by increasing its density. A more efficient core design also allowed the diameter of the spherical containment vessel to be reduced.[24] As at Bradwell, a single massive charging and discharging machine was used.[25]

Fig. 5.16
Dungeness, Kent, April 1963, showing the extensive construction camp and the Goliath crane used for the reactor hall construction.
[© Historic England (Aerofilms Collection), EAW011093.

At Sizewell A (Suffolk), a major step forward in station design was represented by placing the two spherical steel pressure vessels in a single structure (Figs 5.17 and 5.18). This allowed for a more compact station design and for a single charge floor, permitting the two fuelling machines to be used on either reactor, removing the need for a spare machine. Here, and at Dungeness (Kent), the increase in pressure vessel thickness also allowed for a slight reduction in their diameters. In a further attempt to make fuelling more efficient, the fuel charge chutes were left in place, whereas in earlier designs a separate machine was used to insert the chutes before fuelling.

As mentioned above, one method of improving the efficiency of the Magnox stations was through increasing their internal gas pressure. The major constraint to this advancement was the thickness of the steel pressure vessel, which was in turn reliant on developing welding technology required to join ever-thicker vessel plates, and quality assurance techniques to guarantee the work. A major innovation in the penultimate Magnox station built at Oldbury (Gloucestershire), probably inspired by earlier French practice, was the use of prestressed concrete pressure vessels (Fig. 5.19). Prior to its construction, design and testing of one-eighth-scale models had taken place, including destructive testing at AWRE Foulness (Essex). This design also removed the complex civil engineering challenges associated with the handling

Fig. 5.17
Sizewell A, Suffolk. During construction, 28ha was covered by the construction camp of offices, assembly areas, storage and workshops. In comparison, the finished station occupied 8ha. Here the existing 50-year-old woodland was retained to partly screen the station.
[© Historic England Archive (Aerofilms Collection), EAW137831]

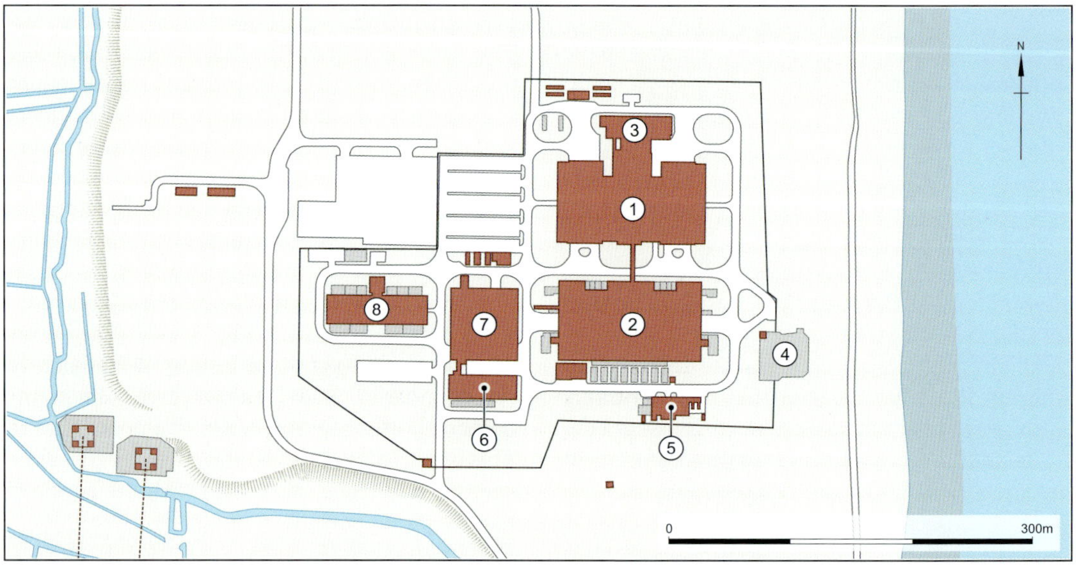

Fig. 5.18
Sizewell A, Suffolk. Diagram showing principal buildings.

Key
1 Reactor building
2 Turbine house
3 Effluent treatment
4 Cooling-water pump house
5 Water treatment
6 Administration and canteen
7 Workshops and garages
8 Switchgear building

[Contains OS mapping. © Crown copyright (and database rights) 2025 OS AC0000815036]

of large steel components, although massive steel heat exchangers were still required. In this design the pressure vessel contained the coolant gas, and housed the reactor, boilers and blowers, as well as acting as a biological shield. The use of a concrete pressure vessel permitted a taller core to be assembled and this, linked with increased gas pressure, led to a reduction in the number of fuel elements required.[26]

Following on from Sizewell A, Oldbury was also designed with a single charge floor, with each reactor served by a single charge and discharge machine.[27] Above the charge floor a travelling crane was able to manoeuvre a heavily

Fig. 5.19
Oldbury, Gloucestershire. To the rear is the reactor building and to its left are the turbine house and covered switchgear building.
[© Historic England Archive, OP04695]

shielded and pressurised reactor servicing machine, manufactured by GEC, Erith, into place to carry out routine maintenance tasks on the rods. Other more specialised tasks might include the recovery of fallen control rods and jammed fuel elements, positioning of fuel elements with thermocouples, insertion of a monitoring camera, and in an emergency the removal of fuel elements. To control these procedures the machines were equipped with television cameras.[28]

Wylfa on Anglesey, the last of the Magnox stations, had a power rating of 1,180MW, which was almost double that of Oldbury (Fig. 5.20). Consent for the station was given in 1962 and it was officially opened a decade later. It too employed prestressed concrete containment vessels, each weighing 81,289 tonnes, holding both the reactors and heat exchangers. To strengthen the vessels, they were formed from 1,338 tendons, representing 4,827km of wire weighing 5,080 tonnes.[29] Following established practice, the station had two reactors with a single charge floor, each served by an individual fuelling machine, whose advanced design permitted the development of smaller units. A single servicing machine was used to maintain the standpipes of both reactors. In contrast to the other Magnox stations, which used cooling ponds to hold spent fuel elements, here they were stored in dry storage cells positioned midway between the two reactors. In these the spent fuel elements were held in tubes filled with dry carbon dioxide which were externally cooled by air rising through the storage cell before being vented up a central stack.[30] One of the disadvantages of having the reactor and boilers contained in the same vessel was access problems. A leak was detected in the boilers shortly after opening, which resulted in a seven-month shutdown.[31]

Controlling the reactor

Following on from the configuration of the early research reactors, the first power reactors were each controlled from a dedicated control room. The linear layouts of the stations at Calder Hall, closely followed by Chapelcross, with relatively widely spaced reactors, also made this the only practical arrangement (Fig. 5.21). At Calder Hall, the control room was located on the second floor, a single control desk facing a series of

Fig. 5.20
Wylfa, Anglesey. The station is dominated by the single grey oblong reactor building. To the top is the cooling-water intake and on the opposite side of the building the turbine house and workshop, and to its left are the administration and canteen buildings. In the foreground is the large, covered switchgear building.
[© Crown copyright. RCAHMW, 6050590]

Fig. 5.21
Chapelcross, Dumfries and Galloway, March 1959. The turbine hall electrical power control room. Instruments on the wall provided information on the four piles, transformers and generators. The room was lit by AEI fluorescent reflector units set behind a square laylight with opal Perspex panels and angled troughing around its edges that directed light onto the control panels.
[Courtesy of the Nuclear Decommissioning Authority, AJ2536]

panels displaying information on temperature, gas and steam pressure, and flow indicators. From here, during operation, the boron steel control rods were gently raised and lowered to ensure a constant output. The control panels also displayed information from the burst cartridge detector room, which was housed above the control room. The core held over 10,000 fuel elements, and one problem that could arise was that a burst fuel cartridge might release fission products into the reactor. To guard against this each fuel channel was fitted with a stainless steel tube connected to the burst cartridge detection room. In this room, machines colloquially known as 'sniffers' were used to detect the gases xenon and krypton, tell-tale indicators of a burst cartridge.[32] In the event of an emergency, a red button on the operator's desk could be used to drop the control rods to shut down the reactor. During its later years the control room was usually staffed by three operators of the same grade as those in the separate turbine hall control room.

At the later commercial stations, a single control room was the norm, usually placed between the reactors or in the turbine hall, which was more economic to construct and staff, and allowed the operators to react more quickly to changing conditions. This room would take over from initial local control of the reactors and turbines once the station was brought on load. To minimise the amount of radiation the staff might be exposed to, the control room was typically placed on the opposite side of the reactor to the cooling pond and spent fuel handling areas. At Berkeley, the station was generally run by the assistant shift engineer from a central control desk, with two operators in front of him to monitor information from the plant.[33] Elsewhere in the plant other control rooms served to carry out specific tasks, such as refuelling and spent fuel handling. An innovation at Oldbury was the use of the world's first computer-based alarm and indication system in a nuclear power station. This was supplied by Associated Electrical Industries Electronics and AEI Automation. The AEI 1040 digital computer was able to monitor 3,000 alarm contacts and was crucial to the safe operation of the reactor, which worked close to its maximum working temperature.[34] This was taken a stage further at Wylfa, which was the first CEGB nuclear station to use a centralised computer system. Two computers were supplied by Marconi, allowing the standby machine to be used for off-line calculations.[35] Careful consideration was given to the design of the control rooms to place the critical instruments on their main consoles. Information on the reactors was displayed on the opposite walls, and other panels showed data on the individual boilers, the gas circuit, its blowers and associated steam-raising plant. To ensure even lighting, most control rooms were illuminated by translucent ceilings, with some directional lighting onto the walls.

Heat exchangers

The heat exchangers, or boilers, were typically produced by established boiler manufacturers (*see* Table 5.1). Two alternative solutions were employed in their manufacture: they might be fabricated in sections and transported to the site for final assembly or, if the station was coastal, manufactured as single units and floated close to the site. For Calder Hall and Chapelcross, Babcock and Wilcox produced the boilers in sections for on-site assembly.[36] At Berkeley, the heat exchangers were manufactured by John Thompson of Wolverhampton, and also moved to site by road for final assembly (Fig. 5.22).[37] At Bradwell, the heat exchangers were floated from Head Wrightson and Company's Thornaby-on-Tees works to a purpose-built landing stage close to the new station, before being hauled into position (Fig. 5.23).[38] At Dungeness, manufacture of its eight boilers was shared between Head Wrightson and Co., who again chose to float fully finished units into place, while John Thompson opted to manufacture their boilers in sections for final on-site assembly.[39] At other locations sea transport was used for most of the journey, with the final section being completed by road. This option was also used at Hinkley Point, where the boilers and reactor sections were manufactured by Babcock and Wilcox of Renfrew, and floated down to a specially constructed wharf capable of handling ships of up to 1,524 tonnes at Combwich (Somerset) on the River Parrett (Fig. 5.24). The last 10km were completed by road, many of which were improved to take the unusual loads. At Sizewell A the boilers were produced at the Dalmuir works of Babcock and Wilcox (West Dunbartonshire) and transported by sea from the Clyde to a specifically equipped wharf at Lowestoft (Suffolk), before the final 38km journey by road.[40] For the manufacture of the boilers for Trawsfynydd, International Combustion of Derby erected a new 107m by 62.5m assembly shop at their Sinfin Lane works. From there sections were taken to Liverpool docks for transfer to a newly built wharf at Porthmadog and finished their journey by lorry.[41]

Fig. 5.22
Berkeley, Gloucestershire. Heat exchangers manufactured by John Thompson, Wolverhampton, being welded together on site in a temporary shed.
[© Historic England Archive (John Laing Photographic Collection), JLP01_01_077_10]

Fig. 5.23
Bradwell, Essex, 19 May 1958. A heat exchanger being transferred to land for the final leg of its journey.
[© Historic England Archive (Aerofilms Collection), EAR31450]

Fig. 5.24
Combwich, Somerset, 30 June 1959, showing the new dock with a derrick crane that was constructed to take delivery of the large components for Hinkley Point.
[Historic England Archive (RAF Photography), raf_58_2984_0124]

The turbine hall

At most of the Magnox stations the building with the largest footprint was the turbine hall, which stood outside the areas controlled for possible radioactive exposure (Fig. 5.25). This might also incorporate the station control room for the reactors and generators, annexes for workshops, auxiliary turbines, water treatment plant, battery rooms, external transformer bays and, in the unlikely event of a power outage, emergency diesel generators to power critical plant.

The turbine halls had a substantial basement with massive concrete foundations to support the heavy turbogenerators. The main superstructure was of steel portal frame construction internally supporting travelling cranes for plant assembly and maintenance. Externally, at ground level they were often clad in brick with metal-framed windows matching that of the remainder of the station. Their expansive upper areas were covered in either asbestos or light metal sheeting, often with large glazed panels to light the interior.

For safety reasons, the turbine generator sets were positioned transverse to the building. Should a catastrophic failure occur in the set, spinning debris would be directed away from the reactor. Besides the early stations at Calder Hall and Chapelcross, there was no consistency in generator output or manufacturer, although most sets were manufactured by C. A. Parsons & Co. Ltd of Newcastle upon Tyne (*see* Table 5.1). The number of turbines varied between stations. At Calder Hall two were installed in each of its turbine halls, while three, four, or six were installed at other stations. Later, for example at Sizewell A, the use of a pair of more powerful turbines led to a smaller building. Not all changes to power ratings left an obvious physical expression, and at Calder Hall and Chapelcross the turbine power was increased from 23MW to 27MW by installing improved blades.[42]

Each station was connected to the National Grid through a switchgear compound. Ideally, on cost grounds these were left exposed, but at coastal locations such as Dungeness, Hunterston and Wylfa, the electrical equipment was enclosed to provide protection from salt deposits that might cause short circuits.

Magnox fuel supply cycle

As discussed in Chapter 3, many types of nuclear fuel were fabricated at Springfields (Lancashire). This included the distinctive natural uranium fuel elements canned in a magnesium alloy, which gave its name to the Magnox fleet of

Fig. 5.25
Bradwell, Essex. The turbine hall with six C. A. Parsons & Co. Ltd turbogenerators, each rated at 52MW.
[Nuclear Power Group 4082, private collection]

reactors (*see* Fig. 5.1). Once loaded, an average fuel element spent around six years in a reactor; however, due to the chemical instability of the cladding, indefinite storage was not an option and the fuel elements required reprocessing. Initially, for Calder Hall, and later Chapelcross, this work was undertaken in the reprocessing plant, B204, originally built for the Windscale Piles. It was recognised that to support the Magnox programme far larger facilities would be needed, and design work began in 1958.[43] Windscale was chosen as the site for the new plant, most of which was constructed alongside the existing building and came into full operation by 1964, when the old B204 facilities closed.

After removal from the reactor, to reduce their radioactivity, the fuel elements were generally stored in a station cooling pond for at least 90 days before being loaded into heavily protected steel transport flasks for movement to Windscale (*see* Fig. B5.2.1). On arrival, the flasks were taken to the Storage and Decanning Plant B30, a large rectangular, anonymous sheet-clad building, its walls punctuated by occasional windows. The fuel elements were removed from the flasks and transferred to an open water-cooling pond, measuring about 149m by 17.4m, where they remained for around 175 days. The water in the pond acted both to remove any residual heat and as a biological shield. In 1974 issues in the reprocessing plant caused the flow of material in the plant to back up, which resulted in the corrosion of some fuel cans and increased radiation levels, with knock-on effects on efficiency lasting for many years. From the pond, the fuel elements were taken to the decanning facility where the magnesium alloy cans were mechanically stripped from the irradiated fuel rods. The resulting swarf was then moved to the adjacent Intermediate Level Waste Storage Facility B38 – another large rectangular building, measuring about 116m by

Fig. 5.26
Windscale, Cumbria, B205 Reprocessing Plant. View of the top of the cells where plutonium was extracted from spent fuels for conversion into plutonium oxide for use in the manufacture of plutonium/uranium fuels, June 1964. [Courtesy of the Nuclear Decommissioning Authority, AJ4184]

Fig. 5.27
Windscale, Cumbria, B205 Reprocessing Plant, control room, June 1965.
[Courtesy of the Nuclear Decommissioning Authority, AJ4374]

15m – where it was stored in underwater silo compartments. The facility originally comprised six compartments, but to meet increasing storage demands it underwent extensions in 1972, 1981 and 1983, resulting in an additional 16 compartments. A hazard in this building was the build-up of hydrogen caused by an exothermic reaction in the magnesium. This was countered by maintaining negative air pressure in the compartments and with an emergency system whereby the air in the compartments could be replaced with inert nitrogen.

In the Reprocessing Plant B205, which opened in 1964 (Figs 5.26 and Fig 5.27), the fuel rods were dissolved in nitric acid to separate the uranium, plutonium and waste products. Typically, irradiated fuel comprised by weight 96 per cent uranium, 1 per cent plutonium and 3 per cent waste. The recovered uranium might be stored, or in the case of foreign customers returned to origin, or sent to Springfields to re-enter the fuel cycle. Similarly, the plutonium might be stored, returned to sender, or used in the manufacture of fuel for the Dounreay fast reactor; the waste, some of which was highly active, passed into long-term storage in stainless steel tanks.[44] B205 finally closed in July 2022, after processing 55,000 tonnes of Magnox fuel, representing around 5 million fuel rods.[45] In addition to reprocessing spent fuel from the home stations, it also handled fuel from the Magnox stations at Latina, Italy, and Tōkai Mura, Japan.

In November 1985, Prime Minister Margaret Thatcher opened the new Fuel Handling Plant at Windscale. Commissioned a year later, its function mirrored the aging B30 plant, which was closed at this point. In the new building, Magnox fuel was received, stored and the magnesium fuel-canning removed, prior to reprocessing. Adjacent to it a storage pond for AGR fuel was built, and also the Site Ion Exchange Plant where radionuclides were removed from the cooling-pond water prior to discharge into the Irish Sea.

Box 5.1 The nuclear supply industry

The development of the nuclear industry was heavily dependent on the world-class capabilities of the United Kingdom's existing chemical and engineering industries. These included Imperial Chemical Industries (ICI), which during the war had developed processes to enrich uranium, and afterwards assisted in the development of factory production flow charts. ICI was, in turn, supported by such companies as Metropolitan-Vickers Electrical Company, of Trafford Park, Manchester, which supplied some of the earliest uranium diffusion machines.[1] The post-war atomic programme placed great strain on industries still recovering from the pressures of wartime production as they faced new demands for novel plant and equipment, often to challenging delivery timetables. Besides the more specialised equipment, vast amounts of standard industrial fittings were needed for the more than 1,000 buildings built during the first decade of the nuclear programme.[2]

Many firms were able to meet orders from their existing premises, such as the large turbine manufacturers Parsons of Newcastle upon Tyne, and English Electric at Stafford. Their products, although similar to those supplied to conventional power stations, often needed to be configured to make the most efficient use of the differing steam pressures and temperatures produced by nuclear reactors.

The nuclear industry also created many new opportunities. At Heaton Chapel, Stockport, Fairey Engineering Ltd, which had its roots in the aviation industry, diversified to produce precision engineering components for the nuclear industry. Among its exports were a training reactor sent to Chile and five training reactors to Brazil.[3] The John Thompson group, Wolverhampton, included boilermakers and a specialist subdivision supplying such items as naval gun directors and rocket launchers, as well as industrial control equipment. It took the opportunity to expand its interests into reactor vessels, associated heat exchangers and fuel-handling equipment, supported by its own research facilities.[4] Metropolitan-Vickers, an established supplier to the industry, built an irradiated materials laboratory in Wythenshawe with dedicated facilities for radioactive test pieces.[5]

To the south of Sheffield at Bradwell, Derbyshire, within the newly designated Peak District National Park, the Newburgh Engineering Company reputedly capitalised on its expertise with high-nickel-content pipework for ice-cream production to transition to the nuclear industry. Manufacturing in the early 1950s was undertaken in existing buildings, but to test long assemblies that might be inserted into cores, such as charge mechanisms and grapples, a dedicated test structure was required. Built in early 1963, its footprint measured 14.33m by 5.18m and its lower portion was 4.57m high, with the test tower rising to six storeys 8.29m high. Test pieces up to 12.8m in length were assembled horizontally in the lower part of the building, and a hipped section of roof then allowed items to be raised to a vertical position in the tower. Within the building, clean conditions were maintained with neoprene seals around the door, and workers entered through a clean condition barrier. The building was also supplied with filtered air at positive pressure. To simulate conditions within a reactor, a pit was placed at the base of the tower to hold a pressure vessel. Given its location within a national park and sensitivities about its appearance, the building was faced in local limestone.[6]

Another small business to seize and benefit from the opportunities presented by the atomic age was 20th Century Electronics Ltd, now Centronic. Originally founded in Kent in the late 1940s by two television pioneers, it moved into radiation detection equipment, including Geiger–Müller tubes, often working in association with UKAEA. Business was brisk, and in 1953 the company could afford to move to a purpose-built factory on a greenfield factory estate in Croydon. Typical of many mid-century bypass factories, it had a double-storey, brick-faced office front with projecting stair turrets with laboratories and workshops to the rear. Throughout the following decade openings arose, including demand for radiation monitors for nuclear power stations, leading to the extension of the factory façade along the main road, to the rear of which a mixture of flat-roofed and north-light-roof workshops was added, with yet further growth in the late 1960s. The most distinct features of the works were two towers, one 15.24m and the other 22.86m, which held distillation columns to produce the isotopes boron-10 and carbon-13; production ceased in 1999. The factory was for some time the world's leading supplier of boron-10, and it remains in business as a leading international supplier of radiation detection and monitoring equipment.[7]

Graphite

The nuclear power programme created a huge call for graphite. Prior to the atomic programmes, one of its main uses was for electrodes produced by companies such as British Acheson Electrodes of Sheffield, which were initially important graphite suppliers to UKAEA. The demand is illustrated by the cores at Berkeley, which were composed of over 93,000 graphite bricks machined into 600 different shapes, with a total weight of 2,032 tonnes. To meet this demand, AEI-John Thompson Nuclear Energy Co. Ltd with Morgan Crucible Co. Ltd formed the Nuclear Graphite Co. Ltd, based at Poole (Dorset). The compound was produced from petroleum coke and other materials bonded by pitch or tar and heated to 1,500 degrees Celsius. This was then extruded into 20.32cm by 20.32cm by 7.62cm blocks, which were heated to 3,000 degrees Celsius in a graphitising kiln, before being machined to size.[8] The resulting graphite had a hardness comparable to an H pencil lead.[9] Other important suppliers included A. Reyrolle & Co., Newcastle upon Tyne, which provided blocks for Bradwell.[10] Fairey Engineering, Stockport, produced the graphite blocks for Trawsfynydd and Dungeness B, as well as for smaller export research reactors.[11] During construction, to prevent the ingress of any neutron-absorbing dirt or dust that might impair the efficiency of the reactor, assembly of the graphite core was carried out under extreme clean conditions. Workers and visitors were required to wear boiler suits (secured by tapes to prevent the loss of buttons), gloves, hats and soft rubber shoes.[12]

Depleted uranium

Depleted uranium, a by-product of reprocessing fuel rods, is a very heavy, dense and hard metal; these and other properties make it attractive for several military applications, including use in armour plating systems, penetration heads on bombs and missiles, and in the manufacture of anti-tank shells. It is also used as balance weights in some types of aircraft and yachts. Another benefit over comparable metals, such as tungsten, is that as a by-product, it is a relatively cheap raw material. It has several distinctive characteristics and requires specialised facilities for its handling and to ensure the safety of the workers during manufacturing. In England, a single factory for the manufacture of depleted uranium shells was set up in 1984 by Royal Ordnance Speciality Metals in a repurposed wartime building at Featherstone (Staffordshire) (Figs B5.1.1 and B5.1.2). In operation, depleted uranium was imported in powder form from the United States; then through various pressing and heating (sintering) processes it was rendered into solid billets before machining and finishing into anti-tank rounds. Production ceased in about 2003.

Between 1960 and 1997 the Atomic Weapons Establishment also maintained an outstation in a former Royal Ordnance Engineering factory at Llanishen, Cardiff, primarily for the specialised manufacture of uranium and beryllium components used in nuclear weapons.

Fig. B5.1.1
Featherstone, Staffordshire, Royal Ordnance Speciality Metals. A secondary roof was put over the wartime building to house the filtration plant.
[© Historic England Archive, AA51029]

Fig. B5.1.2
Featherstone, Staffordshire, Royal Ordnance Speciality Metals, depleted uranium machining areas. This view shows a machine position with under-floor filter pits and ventilation hoods above; the pipes carried coolant and demineralised water for firefighting. The marks on the walls and floor are for sampling during decontamination.
[© Historic England Archive, AA05104]

Box 5.2 Transport

As the nuclear industry grew, so too did its requirements to move increasingly large amounts of potentially dangerous materials between sites. The atomic weapons establishments used relatively small amounts of radioactive raw materials for experiments, trials or as components within weapons. Road transport was adequate for the quantity of supplies being moved and provided greater flexibility than the rail network. The roads also offered sufficient capacity to deliver chemicals to the main research facilities. Elsewhere, existing rail connections were used to deliver bulk chemical supplies needed for fuel fabrication at Springfields (Lancashire) and Windscale (Cumbria), where large quantities of nitric acid and caustic soda liquor (for regulating acidity) were required during reprocessing.[1] Both plants were conveniently located close to their suppliers in the industrial North West.

From the late 1950s, as the first nuclear power programme developed, progressively greater quantities of spent nuclear fuel rods required transporting to Windscale for reprocessing. Rail was the obvious solution for moving these hazardous irradiated rods held within heavily shielded containers. Typically, a railhead was a couple of kilometres from a power station, and a flask of spent fuel would be moved by road before being transhipped from a lorry by an overhead gantry crane onto a specially designed railway wagon (Fig. B5.2.1). The first nuclear materials were probably moved by rail in the late 1950s, with specialised flasks manufactured by Distington Engineering, Chapel Bank, Workington (Cumbria). The flasks were generally a cuboid, single steel casting, with the interior machined out to accommodate metre-long Magnox fuel rods with fins welded onto the exterior to aid the dissipation of heat created by radioactive decay. The strength of this design was spectacularly demonstrated in 1984 when a 140.5-tonne diesel locomotive with three carriages was deliberately crashed into a cannister on a derailed Flatrol wagon.[2] The flask emerged virtually unscathed and is on display at the Heysham power station (Lancashire) visitor centre. Later, various types of flasks were required to handle fuel elements from different reactors, such as the steam-generating heavy-water reactor at Winfrith (Dorset). Flasks were also needed for foreign fuel rods imported for reprocessing. These included light-water reactor fuel rods, up to 5m in length, placed in French-designed cylindrical containers manufactured by Excellox and Distington Engineering.

One of the economic advantages of exporting reactors was the continuing requirement to process spent fuel, in Europe an activity limited to Windscale and the large French facility at Cap de la Hague, Normandy. Spent fuel from the Magnox reactors at Latina, Italy, and Tōkai Mura, Japan, was transported by sea in purpose-built vessels, first into Queens Dock, Whitehaven (Cumbria) and then by road to Windscale. Later, a specially designed facility was built at Barrow Maine, formerly Ramsden Dock, Barrow-in-Furness (Cumbria), from whence containers were moved by rail to Windscale.

At Sellafield, rail has been important in reducing the amount of road traffic. During the construction of THORP in the 1980s, most of the steel rebar and concrete was delivered by rail. Construction material for the nearby engineered waste vaults at the Low Level Waste Repository, Drigg, was also moved by rail, and from 1978 most waste has also been transported there by the same means.[3] Attempts to encourage staff to travel to Sellafield by train have so far been less successful.

Fig. B5.2.1
Example of an FNA flatrol, probably rebuilt at Shildon in the early 1980s with standard two-axle bogies. This type worked until about 1986 when covered wagons became the norm. In the centre is a typical flask.
[National Radiological Protection Board 17012, Contains public sector information licensed under the Open Government Licence v3.0]

EW5538
2 2 TOP
SWL 25

Power station layout

In comparison to contemporary coal-fired power stations with their associated railway sidings, coal stores, ash-handling facilities and cooling towers, the new nuclear stations had relatively discrete ground plans. In common with other industries handling dangerous materials, such as munitions, clear distinctions were made in plant design to separate 'safe areas', and to place barriers and changing facilities to restrict access to potentially hazardous sections.

A typical nuclear power station included a relatively restricted number of structures, comprising the reactor buildings, a turbine hall, in some, cooling towers, cooling water pump houses, a fuel store, irradiated fuel-cooling pond, electrical switchgear compound, administration building, visitor facilities, and ancillary structures (*see*, for example, Figs 5.12 and 5.18). Topography and pre-existing features were significant influences on station layouts; the linear plan at Calder Hall was dictated by its location, while at Chapelcross the alignment of the former airfield runway was an important consideration in its design. At these stations the areas around the reactors with special working controls for nuclear operations were marked by fences. In the later stations more consideration was given to clearly separating areas where hazardous radioactivity might be encountered from the other buildings. At Berkeley there was a clear distinction between the pump houses, turbine hall and workshops and administration building to the north, and the controlled area to the south. The latter was surrounded by a fence and access into it was through a changing room. To service this area carbon dioxide supplies and new fuel elements were brought along the road that separated the two areas and stored in structures accessible from both sides of the fence. Likewise, spent fuel elements sealed in transport containers could be removed without entering the controlled area. This set the pattern for the subsequent stations. The next development in station layout was made possible by placing two reactors in a single structure, which allowed for a smaller site footprint and, as discussed below, the impact of a station on its local landscape could be further reduced by the careful positioning of component structures, cladding materials and external screening.

Construction

The skills needed to deliver the government's programme of nuclear power stations were beyond the capabilities of a single company, and initially five consortia were established: AEI John Thompson Nuclear Energy Co. Ltd; The Nuclear Power Co. Ltd; English Electric/Babcock and Wilcox/Taylor Woodrow; GEC/Simon Carves; and Atomic Power Constructions, bringing together large civil engineering contractors with heavy engineering and electrical companies (*see* Table 5.2). In 1960, with a contraction of the expected programme, they were reduced to three: The Nuclear Power Group; English Electric/Babcock and Wilcox/Taylor Woodrow; and a new consortium, United Power Constructions (*see* Table 5.3).[46] The stations also brought opportunities for established suppliers and new companies to manufacture specialised material and equipment (*see* Box 5.1).

Until the beginning of the first nuclear power programme, on the grounds of national security, detailed knowledge about nuclear technology had largely resided within the government research establishments. To expand the atomic programme, it was recognised that engineers from private industry would need to be trained, which included secondments to Harwell, as well as intensive lecture courses for other groups.[47] This new openness was also evident in press and public open days, resulting in reports in the contemporary trade press on the achievements of the past decade and the exciting prospects for the future.

Another consequence of this programme was the need for the consortia to establish independent research laboratories to solve the complex engineering challenges of reactor design. It also illustrated the need of wider contemporary private industry to invest in technical research, a trend the government was keen to encourage. In 1958, Atomic Power Constructions Ltd opened a laboratory at Heston (Middlesex) close to Heathrow Airport, where later specialised rigs were built to simulate the conditions that fuel elements and assemblies were subject to in a reactor. A special room with 0.61m thick walls was set aside for radiation studies. Support was also provided to UKAEA in the design of the SGHWR at Winfrith (Dorset), with research into the effects of steam erosion on possible fuel canning materials, stainless steel and zirconium.[48] The General Electric Company/Simon Carves Atomic Energy Group set up a nuclear design and research facilities at the Fraser and Chalmers Engineering Works at Erith (Kent). In addition to constructing specialist test rigs, GEC and the Nuclear Division of Elliot Bros Ltd developed an analogue computer for simulating the performance of a nuclear power station.[49] The CEGB also opened its own research laboratory at Berkeley in 1961, a recognition that UKAEA's research agendas might not always meet immediate commercial and operational technical concerns (*see* Fig. 5.8).

For reasons discussed below, most nuclear power stations were sited on greenfield sites in relatively remote coastal locations. Construction was usually split into four main phases: heavy civil engineering work; mechanical work; electrical fitting; and finally commissioning and loading. Although no two projects were identical, the initial groundwork at Berkeley was characteristic of the effort required (Fig. 5.28). When construction began in January 1957, a new 2.4km road was built by the county council from the junction near Berkeley Castle, and two 35kV power lines laid to supply power for the construction work. In the first few weeks 254,000 tonnes of stone were brought in for roads and the site divided into three sections. Across 14.5ha, assembly shops, workshops, administrative buildings, and a temporary town including a cinema, sports room, canteen, public house and other facilities were laid out. The second area contained the reactors, and in the third were the ancillary structures, such as the turbine hall and water-cooling system.

One of the major considerations in the design of a power station was the handling and erection of its heavy components, in particular the reactor vessel and heat exchangers. So critical was this part of the work that at Calder Hall a scale model was built to understand the placing of hawser mountings and cranes. To either side of the reactor buildings and turbine halls (Fig. 5.29) tracks were laid for derrick cranes to lift steel work and move concrete-filled buckets. One had a lifting capacity of 15.24 tonnes for the installation of carbon dioxide valves and pipes. To lift the reactor sections into place, a king post derrick crane, nicknamed the 'Big Stick', was erected, comprising a 27.4m tower supported by steel hawsers through which the king post was winched, giving a total height of 61m. The heat exchangers were raised into place by a pair of 37.8m lattice posts, or gin poles, and steel hawsers.[50]

Fig. 5.28
Berkeley, Gloucestershire. Excavations for reactor bases and the large temporary work camp, 22 May 1957.
[© Historic England Archive (Harold Wingham Collection), haw_9433_09]

Fig. 5.29
Calder Hall, Cumbria. Construction of one of the piles with the rail-mounted Scotch derrick cranes to either side. These were supplied by Butter Cranes, Hillingdon, Middlesex, who supplied cranes for several stations. The Windscale Piles are to the rear.
[© Historic England Archive, G5074_3]

Fig. 5.30
Oldbury, Gloucestershire, 20 March 1963. Initial ground works underway, with a rail-mounted Goliath crane used for moving heavy components.
[© Historic England Archive (Aerofilms Collection), EAW98210]

An alternative to using derrick cranes and mono towers was the use of rail-mounted Goliath cranes (Fig. 5.30). Although these were more expensive to construct, they freed the site from complex networks of guy wires. At Bradwell, the Goliath crane, built by Clyde, Crane and Booth of Leeds, traversed a 366m length of standard gauge rail to either side of the reactor. With a vertical clearance of 42.7m and span of 54.9m, it was able to run over the reactor halls.[51] A slightly larger Goliath crane was used in the construction of Hinkley Point and was later used in the construction of Sizewell A (Fig. 5.31).[52] At Hunterston, the Goliath crane was designed by the Societé Applevage, Paris, and was capable of lifting loads of up to 355.6 tonnes. It was also used to lift 304.8-tonne prefabricated sections into place, dispensing with the need to place heavy shuttering at great height. For smaller loads, four 7.1-tonne derrick cranes were placed around the reactor for lifting concrete skips, shuttering and reinforcing rods.[53] At Berkeley, it was reasoned that most lifting jobs did not exceed 15.2 tonnes, and so to avoid the expense of constructing a Goliath crane they used two 39.6m mono tower cranes, raised the heat exchangers into place using gin poles, and moved the 203-tonne reactor base into place along specially constructed rails (Fig. 5.32). The requirement for internal travelling cranes also influenced the design of reactor and turbine buildings, whose framing needed to be robust enough to carry these loads.

Magnox reactors were designed to run for many decades with little if any major maintenance. However, over time, as unforeseen

Fig. 5.31
Sizewell, Suffolk, 1 September 1964. This Goliath crane, with a span of 76m and a 61m clear height, could span the reactor building and was equipped with two 203-tonne lifting units. It had previously been used in the construction of Hinkley Point. Also in this view are the temporary workshops for component assembly and to the right of the tall reactor building is the turbine house.
[© Historic England Archive (Aerofilms Collection), EAW137836]

Fig 5.32
Berkeley, Gloucestershire. Heat exchanger being lifted into place, 3 April 1958.
[© Historic England Archive (John Laing Photographic Collection), JLP01_01_078_32]

repairs were needed or new equipment introduced, a place was needed to develop, test and rehearse their use. The former turbine hall of the coal-fired power station A at Littlebrook, Dartford (Kent), was an ideal space for the construction of large test rigs (Fig. 5.33). Here the Facility for Reactor Engineering Development (FRED) was established, later known as the Magnox Electric Littlebrook Reactor Inspection and Repair Rehearsal Facility. Today, although virtual reality is routinely used for training and rehearsals, for particularly complex tasks, many associated with decommissioning, physical training rigs may still be built.[54]

Station architecture

Immediately after the Second World War, the Ministry of Fuel and Power embarked on a massive programme of power station construction. The enormous generating capacity of many of these new stations, such as Giles Gilbert Scott's 1947 Bankside (now Tate Modern) and Staythorpe (Nottinghamshire) by Cecil Howitt and Partners, was matched by their equally monumental brick exteriors. Lord Swinton, chairman of the Committee for Allocating Materials, likened them to 'Assyrian temples'.[55] In post-war Britain, however, they were equally perceived as representing a huge drain on scarce construction materials and manpower. It was estimated that Bankside power station was constructed from 4.2 million bricks, and by 1952 generating stations were using 16 per cent of the available steel.[56] The Committee on Power Station Construction was

Fig. 5.33
Littlebrook Power Station, Dartford, Kent, March 2015. Mock-up of the Bradwell reactor dome in the turbine hall of the closed coal-fired Littlebrook A station.
[© Historic England Archive, DP249413]

established that year under the chairmanship of Sir Hugh Beaver to address these issues.[57] Among its concerns were over-engineered buildings that would far outlast the 30-year design life for most generating plant. The Committee believed that the use of brick cladding on aesthetic grounds was no longer defensible and that modern lightweight cladding materials should be used in place of in situ concrete work. Designers were also encouraged to take note of contemporary practice in the United States, with exposed plant in open-air and semi-open-air stations.[58]

For almost the first decade of the atomic age, engineering considerations dominated the design of nuclear installations. This was an era of post-war austerity and utility standards which matched the culture within the industrial division of UKAEA, directed by Christopher Hinton, who carefully guarded his budgets against any unnecessary expenditure. In tension with the desire to create bold visions of a technological future were long-established views on the aesthetic qualities of the British countryside. Since the 1920s, bodies such as the Royal Fine Art Commission and the Council for the Preservation of Rural England represented public concerns about unchecked growth in the countryside and the promotion of good urban design. During the war, British identity reflected through the country's buildings and landscape were evoked as values worth fighting for, and from the 1940s more formal control began to be exerted through an increasing number of planning controls. More specific direction was given to the electricity-generating industry by the 1957 Electricity Act, which required the consideration of amenity, including natural beauty, in any development.[59]

The architecture of the new nuclear power stations was explicitly modern, creating the vision of an industry in the vanguard of a new industrial revolution – the 'atomic age' – with 'the prospect of a source of illimitable power'.[60] Beyond the massive concrete shield walls, the buildings surrounding the reactors and most of the ancillary structures were supported on steel and reinforced concrete frames, and clad in a variety of materials, including brick, local stone, aluminium and asbestos sheeting and glass. In an era emerging from wartime drabness and smoke pollution, colour, used both externally and internally, was an integral part of the design of the new installations. A contemporary journal predicted that the 'nuclear revolution will add a splash of brightness and cheerfulness to the industrial scene'.[61] At night artificial light shone through their often glass-clad walls, emitting the eerie glow of a new technological age in areas of the country where electricity was still a novelty. The design of the nuclear stations also reflected the increasingly close working relationships between engineers, architects and landscape designers.

Despite the lack of a single architectural vision for British nuclear power stations, the open discussion of site locations through public inquiries and the obligations within public bodies to integrate the stations with their landscapes created a distinctive British style.

In contrast to many conventional coal-fired power stations, which were located close to the consumers of their electricity, most nuclear power stations were placed in relatively isolated places. Factors that influenced their location included nervousness about the new technology (hence a low surrounding population density), the presence of ground that could bear the weight of the reactors, access to vast amounts of cooling water, and a focus on areas of the country lacking local fuel sources where generation costs were the greatest. These requirements often placed them in remote and untouched landscapes noted for their natural beauty. In their favour, compared to contemporary coal-fired stations, most nuclear stations had far smaller ground plans and far fewer fuel movements; approximately 1 ton of uranium will provide the same energy as 20,000 tons of coal.[62] But, in common with conventional stations, transmission lines still marched away from them across the landscape. Most stations were coastal and relied on seawater for cooling, so only at the two early stations, Calder Hall and Chapelcross, were cooling towers built. Exceptionally, at Trawsfynydd a man-made lake, created in the 1920s to supply the hydro-electric station at Maentwrog, was used to supply cooling water.

The Magnox stations

Most of the early nuclear civil and military research installations were designed by the Architects Office of the Ministry of Works, often in collaboration with the Industrial Group of the Department of Atomic Energy, although largely erected by consortia of private contractors. For the subsequent nuclear power stations, there was a desire that private companies should acquire the necessary design skills. A variety of approaches were applied to the architectural and landscape design of the new stations. Some consortia relied on the skills of in-house architects, others appointed architectural consultants who also designed the surrounding landscape, while more unusually a landscape architect worked as part of the team (Table 5.4).

Table 5.4 Magnox power stations' lead architects.

Station	*Operational*	*Lead architects(s)*
Calder Hall	1956	R. S. Brocklesby, UKAEA
Chapelcross	1959	R. S. Brocklesby, UKAEA
Berkeley	1962	K. E. Brundle, W. S. Atkins
Bradwell	1962	Maurice H. J. Bebb
Hunterston	1964	Howard V. Lobb & Partners
Hinkley Point A	1965	Frederick Gibberd
Trawsfynydd	1965	Basil Spence
Dungeness A	1965	
Sizewell A	1965	Frederick Gibberd
Oldbury-on-Severn	1967	Maurice H. J. Bebb
Wylfa	1971	Farmer & Dark

There is no one overall architectural style that characterised the Magnox stations. Factors that influenced the final appearance of the stations included regulations imposed for the handling of radioactive materials and considerations regarding plant construction, particularly the handling of heavy reactor components and the heat exchangers.

The two early nuclear power stations at Calder Hall and Chapelcross (*see* Figs 5.3 and 5.7), originally conceived in 1953 and 1955 primarily for plutonium production, may be seen to be part of the first design phase. The asbestos sheeting and glass panel steel-framed reactor buildings show a clear familial relationship to the design of the research establishments. The initial design for Calder Hall was carried out within the Ministry of Works led by Sir Charles Mole, before it was passed to UKAEA's Industrial Group at Risley, under its chief architect R. S. Brocklesby, for completion, which involved the production of over 1,800 drawings.[63] The appearance of this drab, utilitarian structure was brightened by the careful use of colour. Externally the steel supports for the black heat exchangers were colour coded red, yellow, blue and green. The asbestos sheet cladding was left in its natural off-white state and internally the structural steel was finished in grey, contrasting with the primary colours used for the doors. Handrails and the overhead cranes were painted red. The station's turbines, along with their

Fig. 5.34 Berkeley, Gloucestershire, Reactor No. 1, June 1962. Here the central reactor units were each served by eight heat exchangers. [Courtesy of the Nuclear Decommissioning Authority, AJ3906]

associated plant, were also colour coded. One of the few extravagances in the architecture of the plant was the use of local stone in the gable walls of the administration building and the walls flanking the main entrance.

As the programme evolved, legislation demanded greater concern for the visual impact of the commercial stations on their surroundings – concern which was ameliorated by the careful integration of architecture and landscape design. Different design solutions and philosophies were employed to blend these huge buildings into their local landscapes. At Hunterston, where the consulting architects were Howard V. Lobb and Partners, the station was placed on the coastal plain adjacent to the Firth of Clyde with Goldenberry Hill to its rear. This allowed the station to be positioned so that when viewed from Largs to the north the reactor buildings did not project above the skyline.[64] To give lightness and blend the station into its surroundings, the heat exchangers were enclosed in glazed steel-framed structures, although at the public inquiry Lobb conceded that he had produced the design before seeing the site.[65] Other exterior surfaces were neutral and subdued: grey concrete and grey brick facings between concrete frames; Westmorland slate and local sandstone were used for facings, with odd splashes of bright colour.[66] At Bradwell and Dungeness, built by The Nuclear Power Group, a similar approach was adopted where glass-panelled buildings allowed an observer to almost see through the buildings. At Bradwell, designed by Maurice Bebb, careful attention was paid to the facing materials, which included aluminium sheeting that weathered to a dull grey, Leicester Lilac bricks, and exposed natural aggregate in the concrete panels, while its doorways and other openings were painted in maroon and olive greens.[67] Sylvia Crowe, president of the Institute of Landscape Architects from 1957 to 1959, described Bradwell as 'huge, clean, light and floating almost like one of the clouds over the estuary'.[68] To achieve an architectural effect at Hinkley Point, Frederick Gibberd relied on the textures and geometric shapes of the cladding materials to differentiate between different building functions. To soften the corners of the reactor buildings, the heat exchangers were again clad in glass, while the reactor buildings were covered in aluminium sheeting and semi-transparent glass. The turbine hall was faced in contrasting glass and aluminium panels, and buildings in the administration area were covered in coloured enamel sheeting or local stone.[69] In contrast, at Berkeley, where Giles Gilbert Scott was the consultant architect, the view was taken it could not be blended into its surroundings and instead a simple functional design was chosen, with the different elements of the station clearly exposed (Fig. 5.34).[70] However, where concrete cladding was used, this incorporated local stone to assist in blending the station into the landscape.[71]

The landscape of the atomic age

For most of the early industrial age, heavy industry exploited its surroundings with little regard to its visual impact or the pollution it was creating. Prior to the First World War (1914–18) most electricity was generated locally, either by a local authority to serve its community or by an industrial enterprise for its own needs. During the inter-war period there were attempts to make the system more efficient by building larger regional stations and linking these to a national power grid. On 1 April 1948, along with other key industries, the electricity industry was nationalised to create the British Electricity Authority, later the Central Electricity Generating Board. Public ownership brought with it an expectation that development would be for the general good and, as we have seen, environmental aspirations were embodied in the 1957 Electricity Act. Sir William Holford (1907–75), a leading architect and town planner and the industry's architectural consultant since the late 1940s, offered a vision where 'we must use all the devices of design, orderliness and landscape formation to create the landscape of the atomic age'.[72]

Although the footprint of a given nuclear station was relatively small, its effect on its immediate landscape was far wider. Most stations were sited on flood or coastal plains close to sources of cooling water. At Berkeley, 253,847 cu m of earth was excavated for the reactors and turbine hall foundations. This was used to raise the whole area by about 3m and above the flood level. Nearby, at Oldbury, part of the riverbed was excavated to create a storage lake, this material and earth from the foundation trenches being used to raise the site by nearly 4m (Fig. 5.35; *see* Fig. 5.30). A similar approach was taken at Bradwell where its site was also raised.[73] Elsewhere, the opportunity was taken to create new land. At Hinkley Point, the excavated material was placed behind a 762m sea wall, which provided an extra 6ha.[74] Similarly at Hunterston (Ayrshire), nearly an extra 5ha was created by depositing excavation spoil in the Firth of Clyde.[75]

The buildings within the fenced outer compound of a Magnox station occupied a relatively small area, between about 10ha at Bradwell and 17ha at Hinkley Point A, though initially far larger areas were acquired for their construction (at Bradwell 58ha and at Hinkley Point 56ha).[76] Immediately around the stations land was required for temporary site offices, assembly shops, staff amenities and storage areas (*see* Figs 5.16 and 5.28). Temporary settlements for the construction workers might also be built, including accommodation blocks, shops, canteens, cinemas and bars (*see* Fig. 5.28). Ownership of a larger area also gave the opportunity for wider landscaping and the construction of future stations.

Prior to the construction of most stations, local roads were upgraded for construction traffic and later for use by fuel lorries. Railways that might have otherwise been closed were retained for moving nuclear fuel flasks.

At Oldbury, the CEGB appointed Geoffrey Jellicoe as the landscape designer (Fig. 5.35); his practice also designed their grid control centre at Durley Park, Keynsham (Somerset). The new station was to be placed in low-lying agricultural land defined by irregular fields, many following the sinuous lines of medieval strip fields created by ridge and furrow cultivation. Here the issues included the design of the station and the disposal of 1,529,200 cu m of excavated material from the station's foundations and offshore tidal cooling-water reservoir. It was considered that a new hill created from the excavated material would be an unacceptable intrusion into this area, and instead the upcast was spread over the whole of the 57ha site, raising it by about 3.4m. To blend this plateau with the surrounding fields, its edge was angled at 30 degrees, and where possible existing trees were retained to break up the skyline. Across the newly raised area a more rational field pattern was imposed, although to maintain local character post-and-wire fencing with thorn bushes was used. Closer

Fig. 5.35
Oldbury, Gloucestershire. The cooling water reservoir built out into the River Severn held 400 million gallons of water at ebb tide. [© Historic England Archive, 26756_020]

to the station an unclimbable fence was screened by beech hedging on either side, and a chestnut fence was erected along the public footpath. On the river side several clear views were left of the station.[77]

At Sizewell, a large coastal strip was acquired with the potential to build up to three stations (*see* Figs 5.17 and 6.24). Here Frederick Gibberd worked with the landscape and garden architect Peter Youngman. To reduce the visual impact of the station, the main axes of the buildings were set at right-angles to the sea and positioned 50m behind the old cliff line, although, for practical reasons the pumphouse was set forward in a niche in the old cliff. A change in reactor design permitted them to be housed in a single building, and its presence was further reduced by using dull aluminium cladding on the reactor and turbine buildings, the plain concrete finish of the heat exchangers, and on a low ancillary building. To the south two existing blocks of woodland were retained, and further trees planted to conceal the clutter of car parks and smaller structures. The station is approached from the south and, borrowing from formal garden design principles, a feature was made of the three-storey administration building with a lawn and ha-ha to its front, the latter replacing the security fence in this area. The foot of the administration building was adorned with a raised bed of mixed planting. To break the link between the station and its transmission lines, underground cables were dug at considerable expense between the Switchgear Building and the first pylons.[78] At Hinkley Point, Frederick Gibberd also sought to block views of the car parks and low ancillary buildings by tree and scrub planting.

One of the most sensitive locations for a nuclear station was in the Snowdonia National Park at Trawsfynydd in mid-Wales (Fig. 5.36). Despite opposition by the National Parks Commission and others, there was considerable local support for the station, and the inspectors at the public inquiry were greeted by banners proclaiming 'Pylons before poverty' and 'Power lines not lines of unemployed'.[79] Concerns about the location were met through the local planning committee, the appointment of a landscape architect and by reference to the Royal Fine Art Commission.

Here, one of Britain's leading post-war architects, Sir Basil Spence, worked with the contemporary landscape architect Sylvia

Fig. 5.36
Trawsfynydd, Gwynedd. Lake Trawsfynydd, constructed during the 1930s for Maentwrog hydro-electric power station. The main A487 road crosses the bottom left-hand corner of the picture. Woodland has been used to screen the station from view at ground level.
[© Crown copyright. RCAHMW, 6328408]

Fig. 5.37
Trawsfynydd, Gwynedd. The curving approach road, free from kerbs, pavements and roadside lights.
[© Wayne Cocroft]

Crowe. Underpinning her vision was a desire to maintain the local character of rocky outcrops, scrub and woodland to reduce the station's impact on the local area and to create a setting that blended the buildings into their surroundings.[80] For most people the view of the station was from the west and the A487 road. It was hidden by lowering the transmission buildings and forming the excavated material into mounds. To give them extra height they were planted with predominantly native species, although North American species, such as lodgepole and Sitka spruce, were also used. To merge the trees with the building, the planting was run into the station's compound, and where possible mountain herbage was established in the transmission compound to break up its shape. Projecting above these were the simple abstract shapes of the reactor buildings, finished in materials designed to be in sympathy with its natural surroundings. Light pollution was also reduced by lowering the electric lamp standards. Within the compound the construction yard was recontoured to give it a more natural looking profile and the floor of the electricity substation was planted with white clover and fescue. To emphasise the rural nature of the station, it was given a winding approach road, free from kerbs, pavements and lighting (Fig. 5.37). Closer to the station, Crowe created two small garden areas for the benefit of the workers. The first was the Dragon Garden, with a swirling Welsh dragon, designed by Spence and laid out in stone by William Ifor Hughes, surrounded by low scrubs, herbaceous plants and grasses. The other is now called the Dame Sylvia Crowe Garden, a linear garden laid out on the south side of the site (Fig. 5.38).[81] The Welsh heritage body, Cadw, has registered these as historic gardens. During decommissioning, some of the evergreens will be replaced with native broadleaves and former building sites covered by further planting.

In 1961, Crowe was also contracted to work with Sir William Holford on the design of Wylfa nuclear power station (*see* Fig. 5.20). Here she recognised the very high landscape value of the rugged coast, with an uncultivated gorse-covered shore interspersed with a few windswept trees. Instead of hiding the station, she argued it should become a new focus in the landscape. She proposed that the ancillary buildings and car parks should be on the landward side to preserve the view along the coast. To shield the massive switchgear building, 173.74m by 152.4m and standing 24.38m tall, which at the time was thought to be the largest in Europe, tree-covered mounds were built to the north-east and south-east.[82] The north-western approach to the entrance was also hidden by another artificial tree-covered mound. The buildings in grey and slate tones took their colour palette from the local buildings and walls.[83]

Often one of the most contentious issues

Fig. 5.38
Trawsfynydd, Gwynedd. A section of the garden designed by Sylvia Crowe. It is protected as a Registered Garden.
[© Crown copyright. RCAHMW, 6324078]

surrounding any new power station is the location of high-voltage transmission lines and their effect on the surrounding landscape. Holford strongly believed that overhead transmission lines should be kept away from areas of natural beauty or amenity. Elsewhere, their visual presence was lessened by avoiding skylines, the use of woodland to break up their lines, and by siting them away from areas with other power lines.[84] During this period the design of pylons remained functional and utilitarian.

Summary

The first Magnox stations at Calder Hall and Chapelcross were primarily built for military plutonium production and designed within the Ministry of Works and UKAEA. In common with other contemporary UKAEA structures their appearance was utilitarian. The next generation of commercial reactors commissioned by the Central Electricity Generating Board and the South of Scotland Electricity Board was subject to the amenity clauses within the 1957 Electricity Act, and their location and design became matters of greater concern. The power station buildings were no longer just seen as weather shelters for expensive plant, but as monumental features that would act as focal points in the landscape, often for many miles. To meet these concerns, noted post-war architects, such as Basil Spence, Frederick Gibberd and Maurice Bebb, as well as the architectural practices of Howard Lobb & Partners and Farmer & Dark, and the architects' department of W. S. Atkins were brought into work alongside the engineers. Some took responsibility for the design of the adjoining landscape while others collaborated with the leading landscape architects Sylvia Crowe, Geoffrey Jellicoe and Peter Youngman.

One of the criticisms that has been levelled against the British Magnox programme was the lack of standardisation in the design of the stations. This was partly due to Britain's early entry to the field and the very understandable desire to increase the efficiency of successive stations. During the Magnox programme the principal technological concept remained unaltered: they were natural uranium fuelled, graphite moderated, and gas cooled. Across the programme huge progress was made through engineering improvements in the efficiency of the plants, so much so that the last Magnox plant at Wylfa had a nominal output of 1,180MW, compared to just 200MW at the original Calder Hall station. Advances in welding and inspection technologies had allowed the thickness of the steel reactor pressure vessels to be increased, and with it the internal operating pressure; this was improved further with the use of prestressed concrete pressure vessels. Improvements were also made to the fuel-handling machinery; the early stations needed to be shut down for refuelling and required several machines for the operation.

Conversely, the later stations could be refuelled while they remained on load, with all the operations combined in a single machine. By placing two reactors within a single structure, a smaller station was achieved, with the added benefit of reducing construction costs and visual impact. Parallel developments in computer technology also enabled greater automation and more efficient running of the later stations.

The carbon dioxide-cooled reactors represented a distinctly British approach to reactor design. In 1950s America, partly due to her enormous coal and oil reserves, there was relatively little interest in pursuing nuclear power generation technology. Here the emphasis in reactor research was for more efficient plutonium production and, by the early 1950s, submarine propulsion. For the latter, small, pressurised water reactors were developed. Later adapted for electricity generation, this now represents the dominant technology for power production, although an experimental gas-cooled reactor (EGCR) using helium was built in the early 1960s at Oak Ridge (Tennessee). Many of the technological and political factors that faced Britain's early nuclear programme were comparable in France. For her first-generation reactors, natural uranium was the only viable fuel, and comparatively cheap graphite a better option than expensive heavy water; carbon dioxide was also used for cooling. A small reactor to produce military plutonium became operational at Marcoule in 1958 and a civil reactor at Chinon in 1963.[85] Design studies for advanced gas-cooled reactors were also carried out in former West Germany.[86] The British Magnox design did enjoy some export success and stations were built at Latina, Italy, and Tōkai-Mura, Japan. By an as yet unreported route of technological diffusion, a Magnox-type reactor was also constructed at the Yongbyon nuclear complex in North Korea.[87]

In the next chapter we will explore how Magnox technology was developed to create a more powerful series of advanced gas-cooled reactors (AGRs).

Box 5.3 Was Calder Hall the world's first electricity-producing reactor?

In the Cold War battle for technological superiority and the race to develop nuclear energy for peaceful purposes, there are several claimants for the honour of being the first electricity-generating reactor. The United States' Experimental Breeder Reactor 1 near Arco (Idaho) was the first reactor to generate electricity, lighting four 200W light bulbs on 20 December 1951, while the Russians maintain that their 5MW Obninsk reactor, which supplied electricity to the local grid on 26 June 1954, was the world's first civil reactor.[1] Calder Hall, which officially opened on Wednesday 17 October 1956, at 92MW was the first industrial-scale station connected to a national supply grid, although its primary purpose was plutonium production. The United States claims, however, that the Shippingport (Pennsylvania) 60MW reactor, which went online on 2 December 1957, was 'the world's first full-scale central atomic plant devoted exclusively to civilian purposes'.[2] The United Kingdom's first truly commercial stations were Berkeley and Bradwell, which opened in 1962.

6

The second and third nuclear power programmes

Introduction

As we have seen in Chapter 4, by the late 1950s the United Kingdom Atomic Energy Authority (UKAEA) was concentrating its research efforts on four main types of potential future reactor designs: advanced gas-cooled reactors (AGR), fast-breeder reactors (FBR), high-temperature gas-cooled reactors (HTGCR) and steam-generating heavy-water reactors (SGHWR). Some interest was also being shown in the Canadian CANDU heavy-water system. The graphite-moderated, gas-cooled, and enriched uranium-fuelled AGRs, with their direct evolutionary link from BEPO and the Calder Hall-type reactors, were a familiar technology, and there was the expectation that by the late 1960s they would offer the prospect of smaller, cheaper and more efficient next-generation commercial stations. Notably, the AGRs produced steam at a comparable pressure and temperature to that of the large contemporary fossil-fuelled stations, enabling similar generator plant to be employed. The other reactor technologies, although promising, were new territory which would require longer to bring to fruition. At the time of writing, the AGR fleet is beginning to be withdrawn from service and the survival and appearance of various components described below will change as decommissioning and dismantling proceeds.

Windscale Advanced Gas-Cooled Reactor

To test the AGR concept, the UKAEA Windscale and Calder Hall complex, with its experience of operating gas-cooled reactors and supporting infrastructure, was chosen as the site for a pilot plant, which was placed to the north-west of the redundant Pile 1. Once the decision to construct a prototype had been taken, work on the Windscale Advanced Gas-Cooled Reactor (WAGR) proceeded very quickly. The initial design started in August 1957, approval for its main features was given in March 1958, and the first ground was broken in November of that year. The containment building was complete by August 1961 and criticality was achieved on 9 August. Over the next few months, the station was worked up to full power, achieving a net output of 27.3MW(E) by January 1963. Compared to a Magnox station, the operating temperature was raised from about 400 degrees Celsius to 650 degrees Celsius and the pressure of the coolant gas doubled from 20.68 to 41.37 bar.[1] Working with these challenging conditions posed pressing questions about fuel performance, the effect of the higher operating temperatures on graphite, the consequences of the exposure of reactor materials to an atmosphere of carbon dioxide in combination with high temperatures and pressures, and additionally of safety and control issues through a full fuel-element life cycle.[2]

WAGR was a compact experimental design, conceived for flexibility rather than maximum efficiency, with most of its plant supplied by International Combustion Ltd of Derby. It is dominated by the hemispherical dome set on the conical base of the steel reactor containment building (Fig. 6.1). The dome, which measures 40.8m tall with a maximum diameter of 41.1m, contained the reactor and heat exchangers. Its outer face was insulated by a 3.21cm thick layer of polystyrene protected externally by aluminium sheets. An airlock provided access into the building for personnel, goods and fuel assemblies. To its west was the turbine hall, housing an English Electric 33MW turbogenerator and stand-by diesel generators (Figs 6.2 and 6.3). In between the two were the reactor and electrical control rooms, technical equipment, and a change room for the containment building,

Fig. 6.1
Windscale Advanced Gas-Cooled Reactor, Cumbria. The station is dominated by the aluminium-clad containment building constructed by Babcock and Wilcox. To its left is the control room and turbine hall and to the right the fuel-element building with a travelling crane for handling fuel-element flasks.
[Courtesy of the Nuclear Decommissioning Authority, P-63-26504-R_AA011236]

while to the north stood the administration offices. Adjoining the north side of the containment building was the ventilation and cooling building. On its eastern side was the fuel-element building. In here, new fuel elements were received in transit flasks via a travelling crane, and irradiated assemblies were stored in an underground carbon dioxide-cooled storage drum, or 'carousel'. After cooling, they were taken to two shielded caves for metallurgical

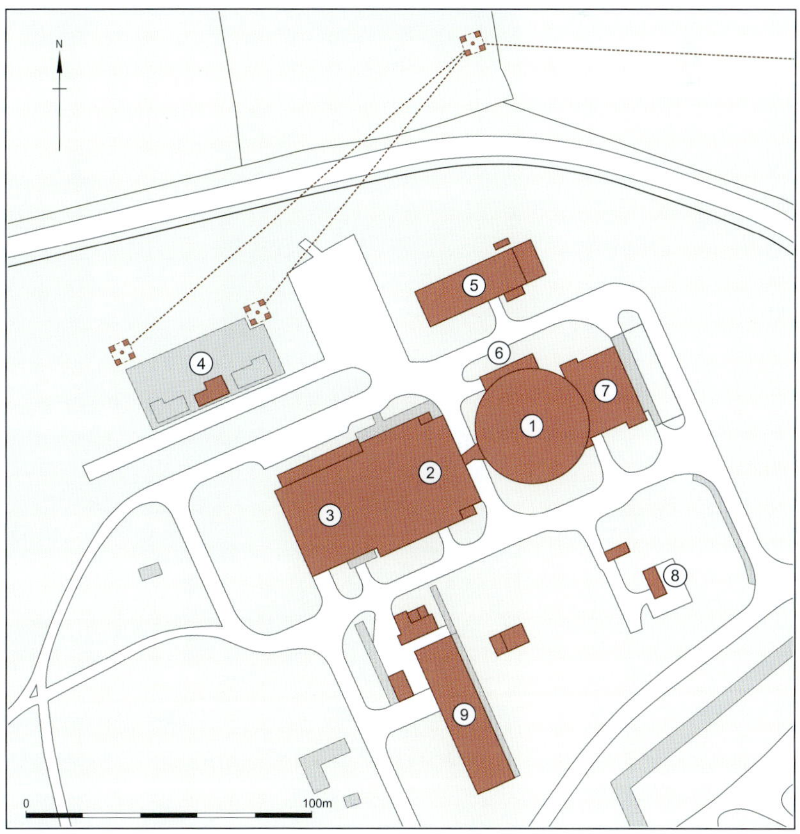

Fig. 6.2
Windscale Advanced Gas-Cooled Reactor, Cumbria, ground plan.

Key
1 Reactor containment building
2 Control and administration building
3 Turbine house
4 Electrical switchgear compound
5 Test loop and electrical plant building
6 Ventilation and air-conditioning building
7 Fuel-element building
8 Gas storage plant and gas discharge treatment plant
9 Forced-draught cooling towers

[Contains OS mapping. © Crown copyright (and database rights) 2025 OS AC0000815036]

Fig. 6.3
Windscale Advanced Gas-Cooled Reactor, Cumbria, pictured in October 1966. The station is dominated by the hemispherical containment building. To its right is the fuel-element building, and to its left the control room and turbine hall. To the right are the four forced-air cooling towers.
[Courtesy of the Nuclear Decommissioning Authority, AJ4628]

inspection and then broken down prior to being loaded into 'coffins' for transfer to the reprocessing plant. To the north of the main group of buildings, and reflecting the experimental purpose of this plant, was the test-loop building for trialling different fuel-element and cooling arrangements. This contained 12 generators to supply the motors of the loop circulators, as well as facilities for carbon dioxide storage.[3] The carbon dioxide and carbon monoxide production plant lay to the south-east, and subterranean plant for treating gaseous effluent before it passed through the 45.72m ventilation stack. To the south-west were four forced-draught cooling towers, and to the north a switchgear compound connecting the station to the National Grid. The layout was designed with enough flexibility that the turbine hall could be extended or another reactor added to the south-west.[4]

The higher operating temperatures of this new reactor design would melt the magnesium canning alloy used in the earlier Magnox fuel elements. To overcome this, stainless steel or beryllium cans were required. However, stainless steel readily absorbs neutrons, hence the need for a more energetic enriched uranium fuel. The departure from natural uranium, used in the Magnox reactors, combined with such elaborate fuel assemblies, was more expensive, but the cost was partly offset by a higher burn-up rate. Enriched uranium dioxide ceramic fuel pellets were loaded into 50.8cm long, 1.01cm internal

Fig. 6.4
Windscale Advanced Gas-Cooled Reactor, Cumbria. Inside the dome two of the concrete biological shield structures for the heat exchangers are visible, and in the foreground, the heavily shielded walls of the fuel-element building. Civil engineering and construction were by Whatling Ltd.
[Courtesy of the Nuclear Decommissioning Authority, P-59-17453-R_AA011222]

Fig. 6.5
Windscale Advanced Gas-Cooled Reactor, Cumbria. One of the four heat exchangers manufactured by International Combustion Ltd, Sinfin Lane, Derby, being lifted into place.
[Courtesy of the Nuclear Decommissioning Authority, P-60_17750-R_AA015930]

Fig. 6.6
Windscale Advanced Gas-Cooled Reactor, Cumbria. View of the interior of the containment building showing the pile cap and to the rear the charge/discharge machine; to either side are the steam-raising units.
[Courtesy of the Nuclear Decommissioning Authority, AJ8500c_AA015924]

diameter stainless steel cans or tubes. Twenty-one of these were then joined to form a cluster, with two clusters in a graphite sleeve forming a fuel-element assembly. Four assemblies were then joined together to create a 4.27m long fuel stringer. This was in turn attached to a plug stringer with an overall length of 17.98m which was loaded into a reactor channel. This arrangement of many smaller cans also provided a large surface area for more efficient heat transfer.[5] After initial experimentation, beryllium was subsequently ruled out as a canning material.

The reactor core comprised a graphite moderator 4.57m in diameter standing 4.27m tall with 253 vertical channels each with a 12.7cm bore. About 200 of these were used by fuel assemblies; the remainder housed control rods and experimental materials, leaving eight channels to be filled with boron stainless steel balls to halt the reaction in case of an emergency. The graphite moderator, manufactured by Whessoe Ltd of Darlington, was held in a double-skin 118-tonne steel pressure vessel, measuring 6.4m in diameter and 16.3m tall. Between the moderator and the reactor vessel was a 15.24cm mild steel thermal shield, the intervening space being cooled by carbon dioxide. Above the core was a shield that prevented neutrons entering the coolant gas ducts and protected the charge floor. After passing through the reactor, the now high-temperature carbon dioxide passed through four heat exchangers. These 190-tonne, 20.4m long and 3.4m diameter units were housed in individual concrete biological containment structures (Figs 6.4 and Fig. 6.5). The reactor was designed to be refuelled on load by a single machine manufactured by John Brown (S.E.N.D.) Ltd. Standing 19.81m high and weighing 406.42 tonnes, this was capable of loading or extracting a complete fuel stringer, control rods and experimental assemblies (Fig. 6.6).[6]

WAGR ran for 18 years until April 1981. During this period it achieved a high availability for research and development and for power generation.[7] It had successfully proven AGR technology, although, with hindsight, a full-scale prototype might have averted some of the technical issues experienced by the following generation of early commercial AGRs. Subsequently, the internal plant of WAGR has been dismantled and the largely stripped structures wait for final demolition.

Fig 6.7
Windscale, Cumbria, HERO reactor in former Pile 1 Blower House B13. [Courtesy of the Nuclear Decommissioning Authority, P-65-7856-1_HERO-0001]

HERO and APEX

In 1959, to support the development of the WAGR, work began on the £1.25 million HERO (Hot Experimental Reactor 0-power) reactor and associated fuel-element examination laboratories in the adjacent and redundant Pile 1 Blower House B13 (Fig. 6.7). HERO was so designed that its graphite core, 5.79m in diameter and of equal height, could be readily altered to study a variety of core geometries and fuel-element designs. This was housed in a stainless steel containment vessel surrounded by a concrete biological shield. In order to dispense with problems of residual radioactivity, it was designed to operate at very low energy levels, with electrically heated carbon dioxide pumped into the core to mimic the conditions within an AGR. It became operational in February 1962.[8] Also built as part of this programme was a far cheaper subcritical facility, APEX, costing £75,000. Both operated for a relatively short time before they were adapted for new experiments in 1962, including the use of plutonium-bearing fuels in HERO.[9]

The second nuclear power programme

The publication of the White Paper *The Second Nuclear Power Programme* in 1964 marked the beginning of the next phase of nuclear power generation. It envisaged that four new stations in England and possibly one in Scotland would become operational by the early 1970s.[10] The type of reactor was not specified and there were differences of opinion between the United Kingdom Atomic Energy Authority (UKAEA) and the Central Electricity Generating Board (CEGB) on the best options. The former advocated for AGRs, sometimes referred to as Mark II gas-cooled reactors, based on the prototype plant at Windscale (Table 6.1). This was also the government's preferred option, as it would ensure the survival of a nuclear supply industry,

Table 6.1 Summary of the AGR programme. All stations have two reactors.

Station	*Build began*	*Operational*	*Power MW*	*Turbines*	*Construction consortium*
Dungeness B	1966	1984–2021	840	C. A. Parsons & Co. Ltd 2 × 660MW	Atomic Power Constructions, Sutton
Hinkley Point B	1967	1976–2023	1,200	Associated Electrical Industries 2 × 660MW	The Nuclear Power Group, Knutsford
Hunterston B	1967	1976–2023	1,290	C. A. Parsons & Co. Ltd 2 × 660MW	The Nuclear Power Group, Knutsford
Hartlepool	1968	1984–	1,180	GEC Turbine Generators 2 × 660MW	British Nuclear Design & Construction, Whetstone
Heysham 1	1970	1984–	1,200	GEC Turbine Generators	British Nuclear Design & Construction, Whetstone
Torness	1979	1987–	1,564	GEC Turbine Generators 2 × 665MW	National Nuclear Corporation SSEB
Heysham 2	1980	1987–	1,344	NEI Parsons 2 × 660MW	National Nuclear Corporation CEGB

Principal source for power outputs: UKAEA n.d. *Project WAGR*, https://arquivo.pt/wayback/20090719150823/http:/www.industcards.com/nuclear-uk.htm (acc. 27 February 2024).

with the possibility of future exports. On the other hand, the CEGB, with earlier concerns about coal supply put aside, wished to acquire and run generating plant as cheaply as possible. To this end it was beginning to show an interest in American light-water pressurised reactors which, if acquired, might put an end to indigenous reactor development and lead to a severe contraction in the supply industry.

Ultimately, the CEGB decided to proceed with the AGR option, justified on cost grounds and the future potential development of AGR technology to deliver greater efficiencies, and ability to win export orders.[11] However, in reality it was a decision partly made on the advice of UKAEA, an organisation heavily invested in AGR technology.[12] The first three stations were located adjacent to existing Magnox stations at Dungeness (Kent), Hinkley Point (Somerset) and Hunterston (North Ayrshire), locations where the CEGB already owned some of the land, with outline planning approval. These coastal locations provided sea water for cooling, common infrastructure – such as roads and high-voltage power line connections – and a local population used to living with atomic energy.

Dungeness

At Dungeness, three consortia submitted designs for an AGR. The Nuclear Power Group proposed a boiling water reactor based on an American General Electric station. English-Electric, Taylor Woodrow and Babcock and Wilcox presented a pressurised water design by the United States company Westinghouse in conjunction with Rolls-Royce. The main contract was won by Atomic Power Constructions Ltd (APC) based in Sutton (Surrey) with a research laboratory at Heston (Middlesex). Some of the features of their proposed AGR included advances in gas coolant pressure and turbine output, and the use of larger fuel-element clusters designed by UKAEA. This design also offered the advantage of on-load refuelling, which it was argued gave the system an economic edge.[13] The station was approved by the Minister of Power in May 1965 (Fig. 6.8), but the project was slow to be completed, and became notorious for technical issues, labour disputes and low productivity, which was compounded by inflation that led to huge cost overruns and delays.

Work started on the site in October 1965, involving a consortium of APC, Fairey Engineering Ltd and International Combustion Ltd, with 36 main contractors, three

Fig. 6.8
Dungeness, Kent, during construction, 6 May 1968. To the right is the circular base of the southern reactor, and to its left is the nine-storey services building and the lower, long turbine hall. To the far left is the pumphouse. To the right is a derrick crane and a smaller Goliath crane.
[© Historic England Archive (Aerofilms Collection), EAW182496]

consultancy firms and another 12 contractors and one consultancy firm employed by CEGB.[14] After a catalogue of problems, the final straw for APC came in 1968 when serious problems with the installation of the containment vessel's steel gas-tight liner led to large financial losses.[15] The project was restructured, and from 1970 it was led by British Nuclear Design and Construction (BNDC), with costs estimated to have risen by £40 million from initial estimates of £89–110 million.[16] By 1972, the estimated cost of the station had increased to £170 million, and by 1978 this had climbed to £344 million, £150 million of which was attributed to inflation. By 1981 the figure stood at £530 million.[17] The original opening was set for 1971, but it was not until 7 p.m. on Wednesday 22 December 1982 that the Nuclear Installations Inspectorate gave permission for the CEGB to start raising power.[18]

Research and development

The basic technology used at Dungeness had been proved by the Windscale AGR, which was acknowledged by a royalty payment to UKAEA. But many practical problems and technical issues remained when scaling up data from an experimental plant to a full-size commercial station.[19] The Heston research laboratories of the United Power Company consortium were used to prove the design of Dungeness B. Tests were conducted on the structure of the prestressed concrete containment vessels, and on the welding techniques for stainless steel thermal insulation panels. Investigations were also undertaken on the core's graphite components, and on associated gas flow and thermal problems. Instrumentation was developed for the experiments and for monitoring and controlling the reactor.[20] This was one of the first major construction projects in the country to employ the state-of-the-art American IBM 7094 for running the computer-aided-design FRAN programme. This was critical to the design of the reactor building. This type of computer was also used to control the contemporary Apollo space programmes.[21] Further analysis was also undertaken using the British Ferranti Atlas and Orion computers using ICT program CS401. More traditional wind tunnel tests were conducted to check and finalise the form of the charge hall.[22]

The graphite cores in the AGRs were subject to irradiation five times as great as in the Magnox reactors, and the gas outlet temperature was 675 degrees Celsius compared to 400 degrees Celsius. It took about two and a half years to refine the core's design and test the design of its 64,000 graphite components to meet these conditions. Prototype blocks were developed at APC's laboratories at Heston, while the graphite was manufactured by the Anglo Great Lakes Ltd, Lemington, Newcastle on Tyne, and machined at Fairey's Stockport works.[23]

A more compact design

The AGR station was far more compact than the adjoining Magnox station. Its two reactors were set closer together at 48.77m compared to 121.92m (Fig. 6.9) and most of the new installation was accommodated within a single structure. The twin reactor building is joined with the tall services building to the south, and to the south again is the turbine hall. Still further south, detached and close to the sea, is the cooling water pump house. Advances in reactor technology allowed the floor area of the new station to be reduced to about 50 per cent that of the A station, yet its output was more than twice as great. One drawback of this elegant design was the difficulty of installing complex plant within a relatively confined area, resulting in increased costs; had more space been allowed it may have resulted in an improved internal layout.[24]

The reactor building

The reactor building, 104m long, 71m wide and 65m high, is set on a continuous concrete raft 101m long, 54.5m wide and 3.4m thick. The main internal elements of the building comprised four circular drums arranged in lower and upper tiers, each pair surrounding a reactor and separated by a reinforced concrete rectangular tower housing the fuel handling unit.[25] The superstructure, clad with sheeting, lit by patent glazing and roofed by insulated protected decking, was supported by rings of steel columns surrounding each reactor. The upper tier is dominated by the charge hall, which comprised the two upper drum elements joined by the central rectangular section, in total 82.3m long, 32.9m wide and 38.1m tall, above which was

Fig. 6.9
Dungeness, Kent, advanced gas-cooled power station. The consulting architects were Howard V. Lobb and Partners.
[© Historic England Archive, 26763_006]

Fig. 6.10
Dungeness, Kent. View of the 1,016-tonne refuelling machine which served both reactors.
[Private collection]

the refuelling machine, carried by a Goliath crane set on longitudinal rails (Fig. 6.10). In the roof of each drum is a 5.08-tonne crane used during construction and, set within the central section, a 50.8-tonne crane used for charge machine maintenance. The lower tier is 104m long, 71m wide and standing 26.8m above ground level. It is independent of the reactors which are covered by two braced tower structures. The central fuel handling building is supported by two portal frames. The use of steel framing in the design was also an important consideration as it allowed for some flexibility for the inevitable modifications to plant designs during construction.

The reactors

The reactors are enclosed within thick pre-stressed concrete containment vessels surrounded by various servicing levels. Each of the pressure vessels weighs 40,000 tonnes and is of post-tensioned prestressed concrete in situ construction. They comprise a cylinder with a total depth of 29.9m with a clear internal height of 17.7m and top and bottom caps with a thickness of 6.1m. In this design, the pressure vessel enclosed the core, boilers and gas circulatory systems. Each also had an internal mild steel gas-tight liner which was water cooled.[26]

The reactors sit on 27.4m diameter and 7m thick base caps. Their membranes, core support and restraint structures, and pressure vessels were prefabricated immediately to the north and then moved into place. To assist in their construction a small Goliath crane was installed in this area (*see* Fig. 6.8). Elsewhere, the construction of this station departed from past practice, and instead of a main Goliath crane, three 20.32-tonne derricks were installed around the reactor building for erecting the steel work and lifting other components into place.

As built, the graphite cores were 9.4m in diameter, standing 8.2m tall, and were contained within an inner pressure vessel. They were pierced by 465 channels, which could be occupied by fuel stringers or control rods. To maintain the operating temperature of the core, about 50 per cent of the carbon dioxide cooling was passed downwards between the outer face of the fuel stringers and the fuel channels, before it joined the remainder of the gas to be circulated past the stringers. It then entered the once-through boilers before being recirculated back to the reactor. At the base of the boilers, the gas circulators were designed as separate cartridges that could be removed for maintenance.

Fuel and fuel handling

The AGRs used a standardised fuel comprising hollow 14.5mm diameter cylindrical ceramic pellets of enriched uranium dioxide contained in stainless steel fuel cans or pins, which were lightly ribbed to improve heat transfer to the coolant gas. The mean heat output of this fuel is about three times that of an equivalent amount of Magnox fuel. Thirty-six fuel pins were fabricated into clusters held in double graphite sleeves supported by stainless steel grids to form fuel elements measuring 1.016m in length. Eight fuel elements were joined together to form a fuel stringer that was loaded as a single unit into a core channel.[27] The degree of fuel enrichment of an element might vary slightly depending on its position in the reactor. Both the fuel and the container are chemically inert in carbon dioxide and therefore safer at higher temperatures than the earlier generation of uranium in Magnox alloy fuel elements.

A distinctive feature of the APC design was the use of a single on-load fuelling machine built by Fairey Engineering (*see* Fig. 6.10). Although this sophisticated machine added to the capital costs, it was destined to reduce operational expenses. The 30.5m tall, 1,016-tonne machine was designed to withdraw a plug unit and attached string of fuel elements with a total length of 22.25m, all the while maintaining the coolant gas flow. Fairey also manufactured the reactor's 960 plug units that retain the fuel elements and control rods that were contained in cylindrical tubes 13m in length and 0.23m in diameter. The control rod plug units were interchangeable with fuel elements, but also contained a 50Hz motor, magnetic clutch, eddy current brake, winch and a counterbalance mechanism.[28] In the event of an emergency, the reactor was equipped with 6 black rods of stainless steel with boron carbide inserts and 21 grey rods of stainless steel. Just a quarter of these rods falling into the channels would ensure the shutdown of the reactor.[29]

The central rectangular fuel handling unit comprised a series of vaults which were arranged vertically and horizontally, with labyrinths around the perimeter. In operation, new fuel elements were received into the fuel preparation room where they were assembled into a stringer with a plug unit attached to the top. From here, the stringer was transferred to the fuelling machine for insertion in the reactor. To defuel the reactor a complete stringer was removed and placed in the irradiated fuel disposal room. At this point it might be dismantled to permit the fuel elements to be rearranged to ensure their most efficient use, reassembled and returned to the reactor. Alternatively, spent elements were passed down an inclined chute on the south side of the building into an acceptance pond before being placed into a container referred to as a 'skip' and onward to the cooling pond and eventually removal offsite for reprocessing.[30]

The services building

To the south of the reactor building is the nine-storey, steel-framed services building. It measures 95m in length, 23m wide and its tallest section is 51m above ground level. Beneath the building is an 8.5m deep basement, which contained the active waste vaults, a water treatment plant, a central cooling pond, standby diesel generator sets, and steam pipes from the reactor building to turbine hall. At ground level there were staff facilities and a further water treatment plant. Rising to 76m above ground level was a reinforced concrete active exhaust stack, or chimney, from the air-conditioning plant room. Internally, this was 3m square with 30cm thick walls, but above roof level it had a circular cross-section to reduce wind resistance. On the top floor were two de-aerator tanks, used to remove dissolved gases in the boiler feed water; this section was enclosed by solar bronze glass panels.

The reactors and generators were operated from a single control room situated in the services building. A novel feature of this station was the increased dependence on computers, with a computerised data processing system. This was used to directly control and monitor the plant, replacing many conventional instruments and alarms, with the data presented on screens, teleprinters, and punched tape. The three Argus 5000 series computers were supplied by Ferranti's Moston works, Manchester. In operation, one was held on stand-by, while the other two covered the control and instrumentation of one or both reactor/generator units.[31]

The turbine hall

The turbine hall is a large single-storey building on the south side of the site closest to the sea. It measures 179m in length and 23m high, with a clear span of 40.6m for the two travelling cranes supported on the structural steel framework. Improved steam output from the reactors required only two C. A. Parsons & Co. Ltd 660MW reheat turbogenerators to do the work of the six less efficient generators in the earlier Magnox stations.[32]

On its south side are four ancillary structures housing generator transformer and services compounds and a water treatment plant room. A notable feature of this building is the innovative roof structure constructed from three-dimensional, welded steel triangular trusses. To ensure the maximum amount of natural lighting, the upper faces of the trusses were covered in patent glazing. This design was created to accommodate an enclosed viewing gallery on its south side overlooking the reactor charge floor, and to meet the operating requirements of the cranes.

Hinkley Point

The second AGR at Hinkley Point (Somerset) occupied a similar coastal position adjacent to the Magnox station, where an 8ha site had outstanding approval for such a development (Fig. 6.11).[33] Here too the new station was able to make use of existing infrastructure, including the two 275kV transmission lines leading to Exeter and Melksham, both designed for upgrading to 400kV.[34]

The contract for the construction of the station was awarded directly to The Nuclear Power Group (TNPG) as it was considered that the other two possible nuclear construction consortia were otherwise engaged and not in a position to tender. Work began on Hinkley Point B on 1 September 1967. Its original cost was estimated at £80–90 million rising to £112 million in following years, and by the mid-1970s to £159 million.[35] In February 1976, although about four years behind schedule, it was the first commercial AGR station to raise power, and its second reactor began production in October of the same year.[36]

As with Dungeness, it was a far more compact design than the adjacent Magnox station, with its reactor and turbine halls housed in a single structure (Fig. 6.12). The main reactor building is oriented north to south and is 119.5m long, 50.6m wide and 36.6m tall. This houses two 630MW AGRs contained in prestressed concrete cylindrical pressure vessels with an internal diameter of 18.9m and 19.4m high, their wall thicknesses varying between 5m and 8.5m. Between the two reactors is a central block containing fuel handling equipment, instrumentation and the control room. To reduce the steam pipe runs, the turbine hall adjoins the western side of the reactor building and the two turbines are placed side by side, oriented east to west with their centre lines aligned on the reactor centres.

This station also marked a move to greater automation. With increased use of data processing equipment, the control room size was kept to 12m by 12m. Its English Electric computer was the first to control a nuclear power station and enabled the process to be run by just three operators.[37] Elliot-Automation, a subsidiary of English Electric, supplied the automatic control and safety system. This relied on three different sets of monitoring equipment and would only activate a shutdown if two or more channels reported a problem.[38] The Power and Marine Division of English Electric supplied an automatic sequence control system to start up and shut down the boiler and turbine auxiliaries. This equipment was designed and manufactured in Kidsgrove (Staffordshire).[39] Its coastal position made it particularly susceptible to the build-up of salt deposits; previously this

Fig. 6.11
Hinkley Point, Somerset, 1968. The B station is in the foreground. Around the perimeter, planting shields the lower buildings from view.
[© Historic England Archive (Aerofilms Collection), EAW189532]

equipment would have needed to be greased or housed in a covered structure. Another innovative feature here was a washing system for its 400kV electrical switching station.[40]

The construction method employed was to assemble the reactor in large sections and then to position them using a 609.6-tonne crane (*see* Fig. 6.11). During construction, a large, temporary clean-conditions building, free from the dust and dirt of a construction site, was built for Head Wrightson to assemble the reactor boiler shield walls.[41] An area was also set aside for Whessoe to manufacture the 227.6-tonne gas baffles (used to separate the coolant gas from the hot gas being passed to the boilers), the 609.6-tonne gas baffle dome, and the 152.4-tonne diagrid support for the core.

Improvements to its design over Dungeness allowed Hinkley Point B's gas pressure to be increased from 33.78 bar to 41.37 bar, and alterations in the design of the gas outflow channel enabled the power of each reactor to be raised by

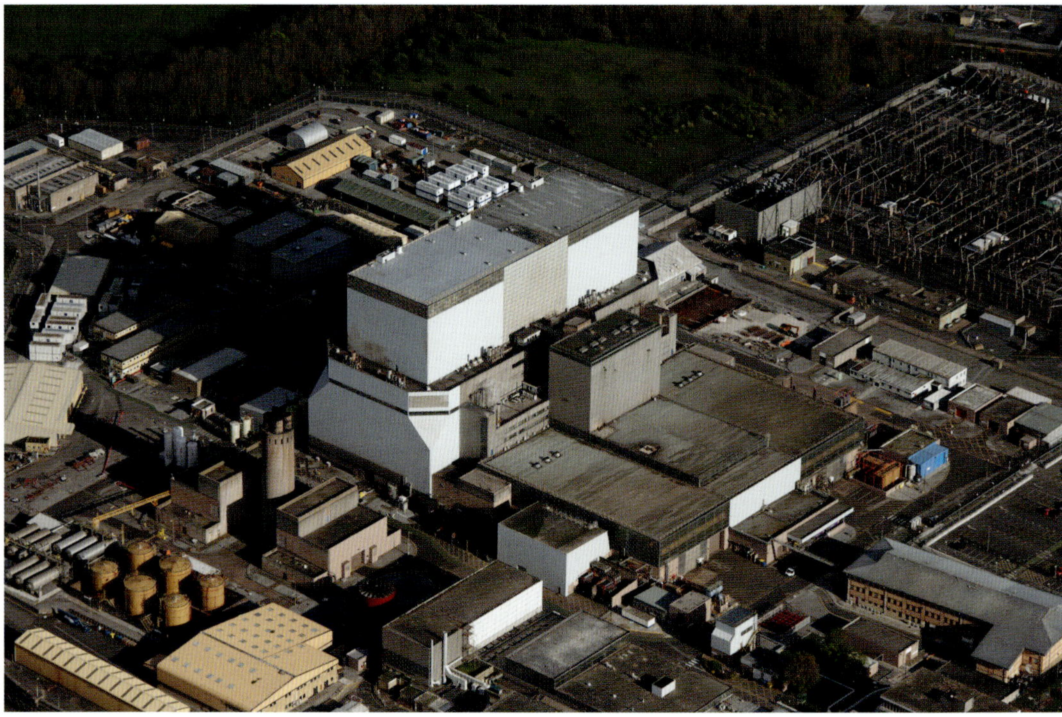

Fig 6.12
Hinkley Point B, Somerset, 15 May 2012. The reactors are housed in the tall white buildings.
[© Historic England Archive, 27472_017]

30MW.[42] The graphite core as built was a 16-sided structure composed of interlocking graphite blocks weighing 1,800 tonnes pierced by 389 vertical channels, 308 for fuel assemblies and 81 for control rods.[43] Each fuel channel had a separate standpipe for loading and unloading. This was undertaken by a single charge machine working inside a common charge hall, which could operate while the reactor was on load. English Electric's Whetstone graphite factory was unable to keep up with the demand for graphite and concentrated on supplies for Hinkley Point B. To supply Hunterston B (*see* p. 211), a new graphite machining factory was established at the Anglos Great Lakes factory at Newburn Haugh (County Durham).[44]

The core is surrounded by a steel liner weighing 457.22 tonnes and measuring 19.35m tall and 18.9m in diameter. It was manufactured in prefabricated panels at Whessoe's Darlington works and brought to site for final fabrication. After completion, it was moved into position along rails by two hydraulic jacks. The liner also acted as the internal shuttering for the prestressed concrete pressure vessel and was fitted with small-bore water pipes to keep it and the concrete cool during operation. Insulation was provided by layers of stainless steel foil and wire mesh wool. The liner also acted to prevent cooling gases leaking through the concrete and transmitted the weight of internal equipment to the outer concrete shell.[45] The shield wall was formed from steel sections containing tubes of calcium hydroxide, which allowed for a thinner wall than in earlier designs. The core and shield, with ancillary equipment, had a combined weight of 3,048 tonnes. These were supported on 16 columns set into the floor of the pressure vessel. In this design, the core and four boilers were contained within the prestressed concrete pressure vessel, a feature that allowed the number of openings to be reduced. The design of the pressure vessel was based on earlier engineering experiences at the Oldbury Magnox station. While the Hinkley Point B pressure vessel was about the same size as Oldbury, the power output was twice as great and the operating pressure far higher. In operation, carbon dioxide entered the boilers at 654 degrees Celsius, which produced steam at 541 degrees Celsius compared to 387 degrees Celsius and 363 degrees Celsius in Hinkley Point A.

Turbines

Electricity was generated by two Associated Electrical Industries (AEI) 660MW turbines. These and associated equipment were manufactured at AEI's plants at Trafford Park and Larne, Northern Ireland. Their installation included a number of innovative features, including the factory pre-assembly of sections of the plant and the use of steel foundations, a feature more commonly used on the Continent.[46] Transformers for the switchgear compound, generator transformers and station transformers were manufactured by GEC/AEI's Wythenshawe Works for installation in 1969; the total value of the contract was over £1 million.[47] To provide emergency back-up, the station was equipped with four Rolls-Royce Olympus gas turbine generators.[48]

Operator training

Initially, the engineers for Hinkley Point B were trained at the University of Bath and, following earlier practice, practical training was provided at UKAEA's Calder Hall Operations School. But in the 1970s, as new types of reactors were introduced, a new Nuclear Power Training Centre was established at Oldbury-on-Severn (Gloucestershire) (*see* Fig. 5.29). This facility included simulators able to mimic the designs of different stations, and test that a trainee was able to operate the reactor in a safe and commercial manner, while introducing complex faults and failures in safety systems.

Hunterston

Invitations to tender for the third AGR station at Hunterston B (North Ayrshire) were sought in summer 1966 from three consortia, Atomic Power Constructions (APC), Nuclear Design and Construction (NDC) and The Nuclear Power Group (TNPG). In October 1967, it was confirmed that the latter had been awarded the contract. This station was to be similarly positioned adjacent to the earlier Magnox station, located at the northern end of a promontory of land facing the island of Great Cumbrane on the Firth of Clyde. This offered an ideal location: a large level area with solid bedrock close to the surface, plentiful cooling water, fresh water for the boilers, and only 56km south-west of Glasgow, its main load centre. It also offered the opportunity to upgrade existing transmission lines to 400kV, without the need to construct additional routes.

Construction

The initial cost of the station was estimated at £50 million, rising in the following year to £88.5 million, and by the mid-1970s to £130 million.[49] Approval to proceed was given by the Secretary of State for Scotland on 28 June 1967. It was envisaged that the station would be operational by 1973, but as with other AGR projects it ran into many delays and finally came on line in 1976.[50]

It was of a similar design to the stations already under construction at Dungeness and Hinkley Point (Fig. 6.13). In common with these stations, it was more compact than the earlier Magnox station, with the reactors, generators and control room housed in a single structure. The reactors were contained in prestressed concrete pressure vessels with integral boilers and gas circulators.[51]

Related to the development of Hunterston B was the establishment of the British Aluminium

Fig. 6.13
Hunterston B, Ayrshire, 1981. The consulting architects were Robert Matthews, Johnson-Marshall & Partners, Edinburgh, and the executive architect E. G. Prior.
[Courtesy of the Nuclear Decommissioning Authority, OP036273]

(BA) smelter at Invergordon. Under the terms of a complex agreement, in 1967 BA had agreed to contribute to the capital costs of Hunterston B in return for reduced electricity costs until 2000. However, escalating electricity costs pending Hunterston B's completion, together with increased contributions to the construction costs, resulted in the closure of the Invergordon smelter in 1981.[52]

The later AGRs – Hartlepool and Heysham 1

The final stations in the second nuclear power programmes, at Hartlepool and Heysham (Lancashire), reflected a greater confidence in the safety of AGR technology, which permitted them to be sited closer to more densely populated urban areas. In the mid-1960s the CEGB had sought permission to construct a fourth AGR station, rated at 1,300MW, on a 48.5ha site which it owned at Seaton Carr, on the mouth of the Tees near Hartlepool. In contrast to other stations, this was an area that lacked obvious visual amenity, characterised by steel and chemical works, and an adjacent shipbreaker's yard (Fig. 6.14).

To date, it was the closest a nuclear power station had been placed to a major urban area, with Hartlepool only 5km to the north.[53] It also marked a move away from areas of the country where coal-fired generation was more costly, and to a region where coal was a major employer. The latter caused much public outcry. The CEGB argued for the need to provide extra generating capacity in the North East, and that base-load generation by a nuclear station would be far cheaper than by a conventional coal-fired station.[54]

The initial cost of the station was estimated at £110 million. Later figures put the cost at £125 million, against £89 million to construct a coal-powered station of similar size. It was, however, estimated to have running costs a third lower; a £37 million saving would be achieved over the lifetime of the nuclear station. The new station was connected to the 400kV network north of Middlesbrough by two short links.[55] In operation the station uses 56,826,125 litres per hour of cooling water. This is supplied by a pump house on Seaton Channel and is returned to the north of Nore Gore breakwater.

Construction work started in November 1968 by a newly created consortium: Babcock English Electric Ltd, comprising Babcock and Wilcox Ltd, English Electric Ltd, Taylor Woodrow Construction Ltd, UKAEA, and the Industrial Reorganisation Corporation.[56] The station was built by the British Nuclear Design and Construction Company, which also later won the contract for the first Heysham station

Fig. 6.14
Hartlepool. Visible are the distinctive buttresses which supported the cranes used in the station's construction. The consulting architects were Frederick Gibberd and Partners. Unusually, this station is located on an urban fringe.
[© Historic England Archive, 28094_070]

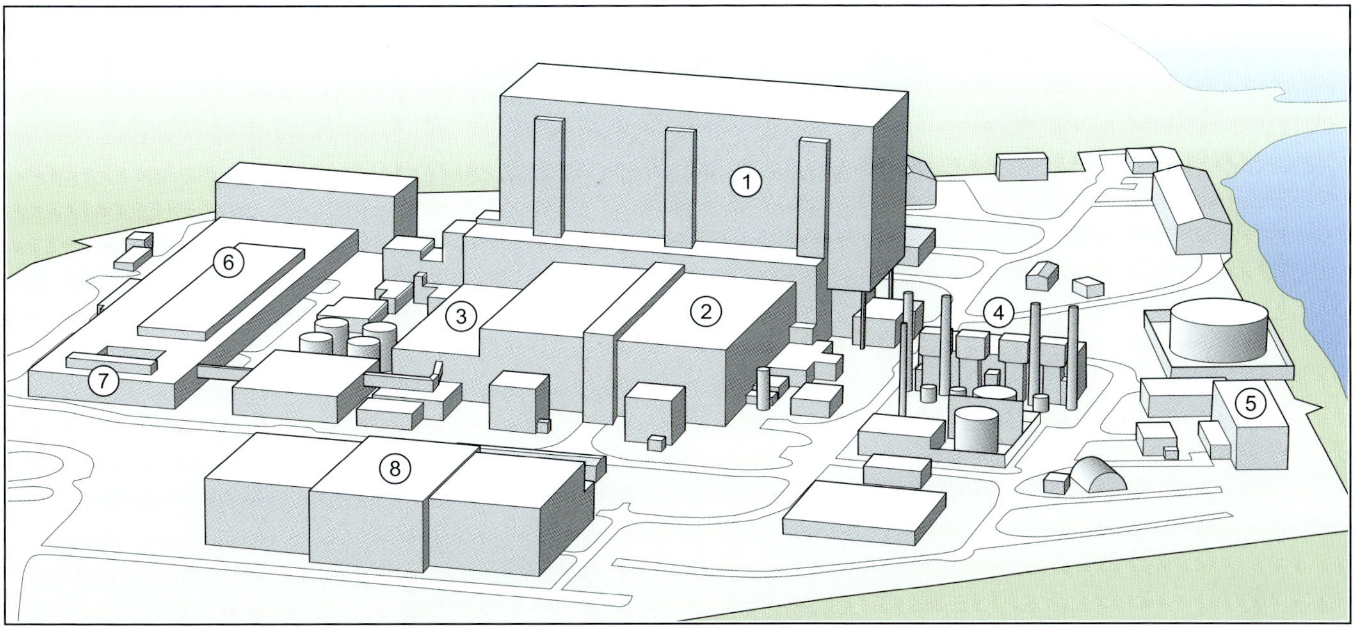

Fig. 6.15
Hartlepool, diagram of main features.

Key
1 Reactor block
2 Turbine house
3 Services block
4 Gas turbine house
5 Cooling-water pump house
6 Stores and workshop
7 Administration and canteen
8 275kV switch house

[© Historic England, drawn from Historic England Archive, 28094_054]

(*see* p. 215). The main contractor for the civil engineering work was Taylor Woodrow. Due to local ground conditions, the foundations of the reactors and surrounding shields were taken down 36.6m on 2.3m concrete shafts that widen out to 3.9m in the bedrock. Other structures were supported on intermediate strata and lighter buildings on concrete rafts.

The initial plan envisaged that the first reactor would become operational in 1973–4 and the second in 1974–5.[57] There were considerable construction delays, and by 1981 the estimated cost had risen to £520 million. Fuel loading began in the first reactor in May 1981.[58]

The design of this station incorporated several innovative features that became standard in later designs. These included the use of eight 'podded boilers' which were placed in 2.7m diameter steel-lined insulated bores, within the 6.4m thick walls of the reactor pressure vessels. These were manufactured in a specially constructed clean facility at the Renfrew works of Babcock and Wilcox and transported by sea by the CEGB's roll-on/roll-off ship *Kingsnorth Fisher*. This both reduced the amount of on-site fabrication and allowed for easier inspection and, if necessary, the removal and replacement of individual boilers.[59] This novel design, however, caused building delays in summer 1970 when the Inspectorate of Nuclear Installations ordered a review of the cap design.[60] The original proposal was for a 127 mm thick 3.4m diameter steel cover; this was later increased to a 2.1m thick prestressed concrete plug.[61]

At the base of each boiler was a compact centrifugal gas circulator that was developed at the Windscale AGR. This feature allowed for the elimination of the large horizontal penetrations that would otherwise have been needed for the circulators. This in turn allowed for a reduction in the internal height of the pressure vessel and a lowering of the core. A thick base cap in the design also permitted the elimination of the need for a core support grid.

Another innovation was the construction of six 61m concrete towers, either side of the reactors, to support the main construction crane, which would also be used for the main structural elements of the reactor building (*see* Fig. 6.14). This system allowed the reactor building to be weather-tight at an early stage, and became a feature of the later four AGR stations, introducing a degree of standardisation in their design which is evident today in the buttress-like features on their side walls. A novel arrangement of the buildings allowed the turbine hall, with two 660MW turbogenerators, to be placed adjacent to the reactor building. This was achieved by siting the common services at the end of the reactor building rather than between the reactors, or between the reactors and the turbine hall (Fig. 6.15). This both saved construction costs and allowed for a reduction in high-pressure pipework and subsequent operating costs.

The core design followed that of the Windscale Advanced Gas-Cooled Reactor. The pressure vessels had an internal diameter of 13.1m and were 18.3m tall. The core was made up of isotropic graphite bricks supported by a system of plates and props. The bricks were octagonal and circular in section with interstitial square blocks. The reactors are fuelled by a single machine travelling on a gantry, supplied by English Electric, which also manufactured most of the other electrical equipment. The refuelling machine held one full-length assembly of eight fuel elements and was designed to operate while the reactor was on load. Irradiated fuel elements were discharged to a cooling pond.[62] Following contemporary trends, the station was computer controlled.[63]

The station's even more compact design gave a cleaner appearance with fewer annexes, although, as with most stations, practical operational needs led to a shanty town of temporary huts and open storage at odds with the original architectural vision. The reactor building is faced in aluminium cladding with patent glazing at high level over the towers. The services building next to the turbine hall is a low building which houses the control room for the reactor and generators, welfare areas and ventilation plant. The turbine hall is flat-roofed, with a deep band of glazing down the centre to

Fig 6.16
Heysham 1, Lancashire, under construction, showing the six towers to support the main construction crane.
[© Historic England Archive (Aerofilms Collection), EAW134482]

provide light to three sides of the building.[64]

In June 1966 it was announced that the CEGB was to begin investigations of reclaimed land adjacent to Heysham docks for the construction of an advanced gas-cooled reactor (Fig. 6.16). No objections on the grounds of amenity were expected. The site offered bedrock comparatively close to the surface and ready access to cooling water. It was argued that the excellent safety characteristics of this type of reactor enabled it to be built closer to urban centres, in this instance just 2km from Heysham with a population of around 17,000. The initial proposal was to build a 2,500MW station comprising four reactors, the first of which would begin operation in 1973 and the final one in 1975.[65] Mainly due to government spending constraints, only a single station was approved in November 1969, at an estimated cost of £72 million.[66]

Work started on this station in December 1970, led by the British Nuclear Design and Construction Company, hence the many similarities to the station at Hartlepool. By this time the estimated cost had risen to about £130 million. Lessons learnt from Hartlepool included the design of 'podded boilers' with concrete plugs from the outset.[67] Its more compact design also placed the joint operating and control services closer to the turbine hall and fuel handling area than in previous designs.

Another common feature shared between Hartlepool and Heysham 1 was the construction of six slightly taller (64m) concrete towers to support the reactor construction cranes and form the main structural elements of the reactor hall. Here they are founded on concrete rafts taken down to rock level beneath the reactor hall basement, and a continuous slip-forming method of construction allowed each to be completed in less than two weeks. The towers are box structures measuring 8.5m by 7.3m, with walls between 0.46m and 0.61m thick.[68] The construction of Heysham 1 was bedevilled by industrial and engineering problems and by 1981 it was five years behind schedule.[69]

The third nuclear power programme

Despite the adoption of AGR technology in the second nuclear power programme, as the 1960s ended the future direction of nuclear power generation technology remained an open question. The choice between the experimental helium-cooled DRAGON reactor, also known as the Mk3 AGR, and the steam-generating heavy-water reactor (SGHWR) was technically and economically finely balanced.[70] It was believed that both were sufficiently well understood for the prototype technology to be transferred to the construction of commercial stations by the mid-1970s. It was also expected that by the end of that decade orders for the first industrial fast-breeder reactors would be approved. Another option was to use light-water pressurised water reactors (PWRs), making use of international, and especially American, research in this field. In France, similar arguments were played out about the desirability of retaining an indigenous gas–graphite capability, the potential of fast-breeder technology, and the probable cost benefits of PWRs. The latter argument won, and in October 1969 Marcel Boiteux, director general of Électricité de France (EDF), effectively announced the end of French gas–graphite development. In the following decade the large expansion of the French nuclear power programme was entirely reliant on American-licenced PWR technology.[71] In Britain, safety fears around PWRs persisted. Lord Christopher Hinton was an especially vocal critic of light-water reactors and in 1974 set out his concerns to the Parliamentary Science and Technology Committee, especially about emergency cooling arrangements should a weld fail in the reactor.[72]

Economically, this period also coincided with stagnant demand for electricity and overcapacity in the generating industry, with little prospect of orders for new power stations until the mid-1970s. This in turn created uncertainty in the nuclear supply industry and its ability to maintain staff and premises.[73] With this prospect of empty order books, the two remaining nuclear station construction consortia, The Nuclear Power Group and British Nuclear Design and Construction, merged to form the National Nuclear Corporation (NNC), with its headquarters at Booth Hall, Knutsford (Cheshire).

Against this background, the trade press reported continuing speculation about searches by the CEGB for new nuclear station sites in Cornwall, Devon, the Solent area, more along the River Severn, and at Connah's Quay in North Wales; none of these schemes went ahead. Despite the flatline call for electricity in the 1970s, the long lead-in times required to design and build new stations meant that consideration was required of power requirements in the 1980s and beyond. Re-emerging threats to oil supplies and the political and social ramifications of miners' strikes which led to the three-day week also made alternative energy sources an attractive option.[74] In August 1973, the CEGB told the Parliamentary Select Committee on Science and Technology that it wished to build 18 new nuclear stations, and that the South of Scotland Electricity Board proposed to build a further eight stations. Unimpressed with cost and construction timetable overruns in the AGR programme, the CEGB had a preference for PWR technology, while north of the border the SGHWR system was favoured.[75] In the following July, Eric Varley, Secretary of State for Energy in the Wilson Labour government, announced the third nuclear power programme would comprise just six SGHWR stations.[76]

The debate over future electricity needs and the most efficient reactor technologies, however, remained unresolved. Further research had allayed some of the safety concerns around PWRs and the first AGRs were entering service. Given that two proven systems were potentially available, government funding was cut to SGHWR research, where considerable uncertainty remained about scaling up the prototype plant. The government's position was finally clarified in January 1978, when Tony Benn, the Energy Secretary, announced approval for two AGRs, and an option for a PWR in the early 1980s.[77]

The two new stations were to be built at Heysham adjacent to the existing plant, and on a greenfield coastal location at Torness (East Lothian). Their design and construction were overseen by the National Nuclear Corporation, and to simplify construction the new station at Heysham was closely modelled on Hinkley Point B, with minor modifications to accommodate improvements in safety features and operational lessons learnt there and at Hunterston B. These included increasing the number of reactor channels from 308 to 332 to achieve greater output, increasing the outlet pressure and heat transfer area in the boiler, and modifying the fuel elements.[78] The expectation was that construction would take about six years at a cost of £900 million, a figure that had risen to £1,430 million by March 1981.[79] Variations in the design away from that of Hinkley Point B, in particular the size of the core and the number of fuel channels, led to delays while the safety cases for the new AGRs were revised.[80] It was estimated that the two stations would each take about 150,000 man-years of work to complete.[81]

Previously, in 1968, approval was given for a new four-reactor station on reclaimed land adjacent to the docks at Heysham, and this including design endorsement from the Royal Fine Art Commission (RFAC), but in the event only one two-reactor station went ahead. In the following decade low growth in the demand for electricity and continuing debate about reactor types delayed any decision about the second station until March 1979, when the RFAC agreed the building and landscape design. Design modifications required for safety and operational improvements led to the need for a larger building, although its height was identical to that of the first station (Fig. 6.18).

The core comprises layers of interlocking graphite blocks. This is pierced by 332 fuel channels holding 2,656 fuel elements, with 89 control

Fig. 6.17
Heysham, Lancashire. To the rear is Heysham 1 and to the right Heysham 2. Of the latter the Royal Fine Art Commission commented that the design 'showed promise of a handsome building'.
[© Historic England Archive, 20319_015]

Fig. 6.18
Heysham, Lancashire, the second station under construction. To the left is the turbine hall; in the centre, white cladding is being wrapped around the charge hall.
[Author's collection]

rod channels. Surrounding the core is a domed cylindrical manganese steel vessel, known as the gas baffle. Each of the reactors is cooled by eight electrically driven gas circulators.

The core, primary coolant circuit and boilers are located within a prestressed concrete pressure vessel, cast around a 13mm thick carbon steel liner. This has an internal diameter of 20.25m and an internal height of 21.87m.[82] It acts as a gas-tight membrane, and all steel tubes that penetrate the pressure vessel are welded to it. It is cooled by the internal gases and on the concrete side by a pressure vessel cooling-water system. The pressure vessel is about 5m thick and acts as the biological shield around the reactor. The boilers are of a 'once through' design and drive two turbine generator units rated at 670MW, manufactured by NEI Parsons, which also supplied the boiler feed pumps and condensing and feed-heating plant.[83]

Running almost in parallel with the construction of Heysham, work began on Torness power station (East Lothian) in February 1979. The station is situated on the Firth of Forth 7km east of Dunbar. This location provided the necessary ground loading for the heavy reactor buildings, cooling water for the turbines, and is reasonably close to Edinburgh, which the station's capacity is comfortably able to supply (Fig. 6.19). Initial activity included opening of Skatelow quarry to provide stone for sea defences. To the rear of this, about a quarter of the 80ha site comprises reclaimed land, using 1.5 million tonnes of excavated material from levelling and excavations for the foundations.[84] Other material was used for landscaping around the site. A purpose-built concrete factory was set up on site to manufacture the 17,500 dolos, or concrete tetrahedrons, used in the construction of the sea defences. The cost of the station was £1,800 million, and the work consumed 1.3 million tonnes of concrete and 10,500 tonnes of structural steel.[85]

In common with the earlier generation of

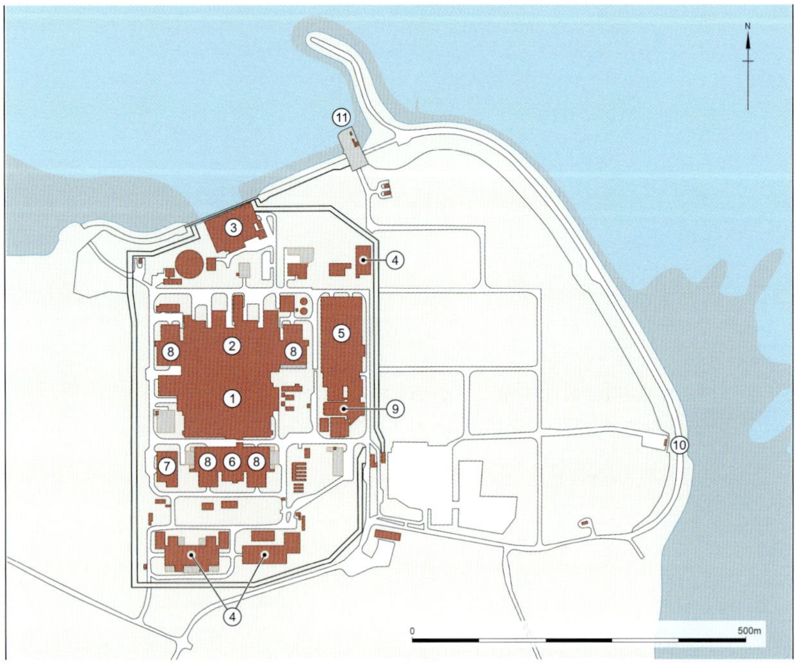

Fig. 6.19
Torness, East Lothian. Diagram showing principal buildings.

Key
1 Reactor building
2 Turbine house
3 Cooling-water pumphouse
4 Substation
5 Workshop and stores
6 Active effluent treatment plant
7 Active solid waste building
8 Diesel generator and switchgear building
9 Administration and welfare
10 Cooling-water outfall
11 Ro-Ro berth

[Contains OS mapping. © Crown copyright (and database rights) 2024 OS AC0000815036]

power stations, the construction of Torness created employment opportunities across the country and a legacy of extended factories. In Middlesbrough, Whessoe constructed a purpose-built factory, Dock Point Riverside, on the River Tees, and refurbished its factories in Darlington for the construction of the reactor liner walls, gas baffles and roof assemblies. Also, in Darlington, Darchem set up a high-tech production facility for the manufacture of 600,000 components required for the pressure vessel's thermal shield. The newly established North Engineering Industries (NEI) group, formed from the amalgamation of several industrial power plant manufacturers, was also an important supplier. In Gateshead, it established a computer-controlled assembly line for the manufacture of the Torness and Heysham 2 boilers. NEI Parsons Peebles in Edinburgh produced the transformers, and NEI Belmos (Lanarkshire) the switchgear. Babcock Power modernised their Renfrew facility for the manufacture of boiler components and supports. GEC Turbine Generators won the contract for two 660MW units, together with the condensing and feed-heating plant, and boiler-feed pumps; work was split between factories in Rugby, Manchester, Stafford and Larne. The diesel generators were manufactured by Mirrlees, Hazel Grove, Stockport, and the boiler-feed pumps by Weirs, Glasgow. The gas circulator motors were produced by Lawrence Scott Electromotors, Norwich. The refuelling machine was manufactured by Strachan and Henshaw, Bristol, where the huge 32m tall, 1,200-tonne machine was built and then dismantled for shipping. Around 2,000 tonnes of graphite was produced by Union Carbide at Sheffield, which was then taken to Fairey Engineering, Stockport, for machining into reactor bricks. Also, in the north-west of England, Ferranti Computer Systems, Wythenshawe, developed the computer control equipment for the station.[86]

To speed up construction, as many of the components as possible were delivered to the site as completed units. It had good links to rail and road and its coastal location allowed the construction of a barge dock for moving the heaviest components, such as the 1,000-tonne roof liner assembly and the 900-tonne gas baffle, which were brought to the site by barge from Teesside. Berthing facilities were also provided for loads of between 250 and 400 tonnes, including transformers, generator sections and turbines.

Over 6,000 people were employed on the project, most accommodated in a temporary village at Thurston Gardens, near Innerwick, which provided accommodation in individual bedrooms for 650 manual and supervisory staff. A landscaped caravan park was built for 37 families, complete with shops and recreational, medical and sports facilities. At the end of the project the sports hall was moved to Yoker, Glasgow.[87]

In common with Heysham 2, the design of the station was influenced by Hunterston B and Hinkley Point B. The main reactor building is 74m high and constructed from reinforced concrete, topped by glazed, steel portal frame structures that cover the charge hall above the reactors. The station comprises two advanced gas-cooled reactors, each weighing 1,800 tonnes, separated by a central fuel-handling building and control building. The reactors are serviced by a single 32m high refuelling machine weighing 1,200 tonnes. This is shielded by higher-density concrete using iron shot in its aggregate and comprises three turrets, used for withdrawing used fuel pins, carrying new pins, and a spare plug unit, used to close the standpipes that provide access to the fuel and control rod channels. Five fuel channels are replaced per month. Spent fuel rods are initially transferred to a pressurised buffer unit for 28 days before being transferred to a decay storage tub and then on to the fuel stringer dismantling facility. In the past the fuel elements were stored underwater for a further 80 days before being placed in transport flasks for transfer by rail for reprocessing at Sellafield.

The reactors are 16-sided, built up from 25,000 graphite blocks and pierced by 332 vertical fuel channels, and surrounded by a 900-tonne gas baffle.[88] Each is contained within a pre-stressed concrete pressure vessel, built up from 3,744 steel tendons formed from seven 18mm steel wires threaded through 8cm diameter mild steel tubes. The walls are 5.8m thick, the top cap 5.5m and the base cap 7.5m.[89] Each pressure vessel is fitted with a mild steel liner, which also serves as a gas-tight membrane. Externally, this liner is cooled by water pipes, and on the inside it is insulated with a Kaowool fire clay fibre blanket, stainless steel foil and stainless steel cover plates. In operation, 8,800kg of high-pressure carbon dioxide is pumped through the eight gas circulators per second. These in turn heat water to 541 degrees Celsius in the four boiler quadrants surrounding each reactor. The resulting steam is fed to the turbines, divided into high-, intermediate- and low-pressure cylinders, with various reheat features arranged to extract the maximum energy.

Immediately to the north are the mechanical and control annexes, with the fuel-handling building and the control room placed between the two reactors. A simulator was also provided for staff training. In common with Heysham 2, to provide security against accidental loss, the four essential supplies buildings, holding the auxiliary Mirrlees diesel generators and switchgear, are situated at the four corners of the reactor building. Adjoining these buildings is the power plant building that houses the 660MW turbines. To the north of the reactor and turbine hall buildings is the circulating water pump house, with the four main circulating water pumps and eight reactor cooling pumps. Between it and the sea is an area with four drum screens, which filter the sea water before it is passed to the pumps.

Thermal Oxide Reprocessing Plant (THORP), Sellafield

The AGRs, and other types of pressurised water and boiling water reactors elsewhere in the world, all use slightly enriched uranium oxide fuels, which require different reprocessing procedures to the Magnox fuel elements. The redundant Windscale B204 plant, renamed the head end plant, was initially used from 1968 to reprocess spent AGR fuel elements, where they were cropped and their contents dissolved in nitric acid before the resulting liquor was transferred to the new B205 building for chemical separation. The remaining stainless steel from the fuel pins was sent for scrap. B204 remained operational until 1973 when a blow-back accident resulted in the contamination of 35 workers, and consequential closure of the plant.[90]

In 1974, following the loss of B204, with more AGRs coming into service and the prospect of international orders for reprocessing, British Nuclear Fuels Limited (BNFL) put forward plans for a massive reprocessing centre, the Thermal Oxide Reprocessing Plant (THORP). This would bring all the processes together in a single structure and, in line with the Non-Proliferation of Nuclear Weapons Treaty, could offer its reprocessing services to non-nuclear weapons states to prevent them acquiring plutonium through the reprocessing of spent fuel elements. Initially, half its capacity was allocated to Japanese customers, who also provided half the capital costs of the plant.[91] Another factor in the plant's design was the need for flexibility to process stored legacy material, comprising a variety of fuels and can shapes. Doubts, however, remained about the wisdom of reprocessing fuels, especially as anxiety about the supply and rising costs of fresh uranium appeared to be unfounded. There were also concerns that reprocessing was adding to the world's stockpile of plutonium, only limited amounts of which were required to produce mixed oxide (MOX) fuel pellets, while the introduction of commercial plutonium-burning fast-breeder reactors was at best sometime off in the future. The United States was especially concerned about the accumulation of plutonium stocks, and in the late 1970s ceased reprocessing and opted for storage of spent fuels until a long-term disposal strategy can be agreed.

Despite these worries, approval for the construction of THORP was granted by Tony Benn, Secretary of State for Energy, in March 1976. Issues surrounding the construction of the plant were examined by a public inquiry in 1977 and, after debate in the House of Commons, a Special Development Order was granted in 1978. Planning for the building commenced in 1981 and construction work began in 1983 on what was, at the time, one of the most expensive and complex civil engineering projects in Europe (Fig. 6.20). Cost estimates for the building

Fig. 6.20
Sellafield, Cumbria, THORP. At the far end of the building was the fuel receipt and storage section; the central block housed the head end plant with shear caves and dissolving tanks; the taller white section was for reagents, utilities and ventilation; the near part was devoted to chemical separation. To the right was the MOX plant.
[© Historic England Archive, 28127_062]

varied between £1.8 and £2.3 billion, slightly more than the contemporary plutonium processing building at Aldermaston (*see* p. 56).[92] Fitting out of the building began in 1992 and the first fuel elements from Heysham entered its shear cave, where they were mechanically cut into pieces ready for reprocessing, in the early hours of Sunday 27 March 1994.

The initial concept for the building was produced by BNFL Engineering following designs by Allott & Lomax Consulting Engineers. Construction was by Sir Robert McAlpine Ltd, with steelwork by Octavius Atkinson Ltd (receipt building) and W. Hare Ltd (pond hall). Shepley Engineering Ltd installed the specialist membranes.[93]

In 2005 THORP was forced to shut down when it was discovered that over a nine-month period 22 tonnes of highly active liquor had leaked from a fractured pipe in an accountancy tank, used to calculate the weight of materials moving through the plant. Operations, although at a lower level, were resumed in 2008, and the plant ran until 14 November 2018. Since that date its ponds have been used to store spent AGR fuel until a permanent solution can be found for nuclear waste.[94] Meanwhile, the remainder of the plant entered its post-operative clean-out (POCO) phase.

THORP is the largest building on the Sellafield site, measuring just under half a kilometre in length and dominated by a 125m tall discharge stack. Externally, it appears as a series of large conjoined, flat-roofed sheds. At ground level, visible to personnel going about their daily tasks, the walls are clad in buff-coloured brick, while the upper sections are covered in brown pressed-metal sheeting, with the projecting staircases picked out in red. The western, pond, end of the building, originally clad in thin aluminium sheeting, has been re-covered in grey and cream-coloured insulated steel sheeting to extend its life.[95] THORP was specified to a very high design standard to contain radiation hazards and to withstand exceptional seismic shocks, as well as more predictable wind and temperature changes. Compared to earlier plants, greater consideration was given to later decommissioning, with extensive use of stainless steel linings in the process areas. To give a 40-year maintenance-free design life, the 192m long pond hall was covered, the trusses, purlins and roof covering also of stainless steel. The 40m roof span is far longer than most available stainless steel members, and the trusses are formed of shorter sections.[96] In the earlier facilities, storage ponds were open, with algae, guano and dead birds adding to the potentially contaminated sludge.

In operation, THORP was divided into three

Fig. 6.21
Sellafield, Cumbria, THORP interior, showing storage racks in the cooling pond before it was filled with water.
[Courtesy of the Nuclear Decommissioning Authority, P-88-1413 NF65_AA010936]

Fig. 6.22
Sellafield, Cumbria, THORP, shear cave. Note the extensive use of stainless steel.
[Courtesy of the Nuclear Decommissioning Authority, P-92-2082-NF6_AA012364]

main sections – receipt and storage, the head end plant, and chemical separation – with the process flowing from south to north. Spent fuel was received in flasks from the power stations: square flasks weighing about 50 tonnes from UK sites and circular ones weighing about 100 tonnes from abroad. Once inside the building the flasks were transferred by a 150-tonne gantry crane running the length of the receipt building to the 14m deep stainless steel-lined inlet pond. This was sufficiently deep for the fuel elements to be removed from the flask underwater before being transferred via a water-filled channel to containers within the main storage ponds (Fig. 6.21). These comprise two large rectangular ponds, whose side walls also support the pond-handling machines used to move the fuel elements. To the south, and detached from the main building, a bank of fan-assisted cooling towers drew heat from the ponds. After a cooling down period, the fuel elements were moved along another water-filled channel to the head end section. Here they were transferred up a ramp to the stainless steel-lined, remotely operated shear cave, where they were chopped into 5cm long pieces before being immersed in hot nitric acid (Fig. 6.22). The resulting liquor was then spun in a centrifuge to remove any fine metallic waste before being transferred to the adjoining chemical separation plant. Here the uranium and plutonium oxides and waste products were recovered. The discarded metal casings were encapsulated in grout and moved to the high-level waste store. During normal operations this massive building appeared deserted, with a staff of only around 20. It was also a largely silent working environment, with little noise except for the constant tick of the criticality alarm system.[97]

Sellafield Mixed Oxide Plant (SMP)

Part of the business case for THORP was that the recovered uranium and plutonium oxides could be used to manufacture mixed oxide (MOX) reactor fuels. It was a particularly attractive proposition as a means of handling recovered plutonium, which otherwise would need to be placed into long-term secure storage. Security concerns about returning plutonium oxide to countries of origin also favoured embedding it with MOX fuel.

Mixed oxide fuels had been produced at Sellafield since about 1963, for use in the Dounreay fast-breeder reactors and to fuel submarine reactors. In the early 1990s, to expand this business the MOX Demonstration Facility (MDF) was built, and from 1993 fuel assemblies were produced for Germany, Switzerland and Japan, which also enabled plutonium to be returned to its place of origin. However, in 1999, falsification of quality assurance records relating to a Japanese order led to large financial penalties, and in the following year a report by the Nuclear Installations Inspectorate resulted in MDF's permanent closure.

Meanwhile, at a cost of around £470 million, alongside THORP, work had begun on the far larger Sellafield MOX Plant (SMP), with a design capacity to process about 120 tonnes per year including use of the plutonium oxide recovered in THORP. The 2005 shutdown of THORP presented a major blow to this model. To meet its contracts, BNFL was forced to subcontract work to European rivals, and also to use plutonium from United Kingdom stocks, which would later be replenished from the processing of foreign fuels in THORP.[98] Externally, the building revealed little of its functions, comprising a series of large sheds, clad at ground level in buff-coloured brick with a grey-coloured metal sheeting central section and a wide white upper band. Again, the external projecting staircases are coloured in red (*see* Fig. 6.20).

AGR station architecture and landscape

The first three AGRs, at Dungeness, Hunterston and Hinkley Point, were sited in locations with existing Magnox stations, and the mitigation of their impact on their local landscapes essentially followed that of their predecessors. At Dungeness, in order to retain the character of the shingle belt, the Royal Fine Art Commission, Nature Conservancy Board Council, and other interested parties requested there should be no landscaping or tree planting around the station. Furthermore, any open areas within the station were covered in shingle. The area's importance as a migratory route for birds was recognised, and during construction a warden was employed to recover any scraps of food left by the workmen to discourage birds from alighting.[99] In contrast, at Hinkley Point B extensive use was made of belts of screening trees to obscure views of the station, leaving only the main reactor blocks and their upper glazing strip visible above the tree line.

The architecture of early AGR reactor buildings may be seen to have a direct lineage from the last Magnox stations at Oldbury and Wylfa, where in place of steel, prestressed concrete was used for the pressure vessels. This permitted a higher working pressure and a more compact design with the reactor, primary coolant circuit and boilers housed within the pressure vessel. The reactors were housed in a single building with a shared charge floor and fuel handling facilities. The upper parts of the buildings effectively acted as weather shelters for the charge floor, although their roof frames might also carry maintenance gantry cranes.

At Dungeness the main reactor building had the dominating appearance of a medieval bastioned gatehouse, its height emphasised by the vertical windows set in aluminium glazing bars. The lower tier of this structure, which is continued through the service building, reflected the height of the prestressed concrete containment vessels surrounding the reactors. Above the reactors the second tier acted as a shelter for the tall charge machine that serviced the reactors (*see* Fig. 6.9). Mounting this machine on a travelling gantry crane that ran on rails allowed a lighter upper building to be constructed than in some of the earlier stations. To the rear of the service building was a taller tower, its upper glazing mimicking that of the earlier Magnox station.

The external base of the structure is finished in brick, and the steel framing above clad in protected steel sheeting. The height of the building was emphasised by vertical windows set in aluminium glazing bars. To cut thermal gain in an already warm building, these were kept to a minimum and glazed in solar bronze glass, which also reduced glare. The glazing also had the effect of reducing the overall bulk of the station by exposing the colours of the plant. Due to its coastal location, salt spray threatened to obscure these windows, and to prevent this, water was directed across the glazing from the roof.

Speaking in 1972, Howard Lobb, the consulting architect, discussed how careful consideration was given to how it would relate to the earlier Magnox station. This was achieved via a visual connection, with the lower level of the new building being finished in matching black and white cladding. Otherwise, black and grey seemed to be appropriate to the seaside atmosphere, and strong bands provided a contrast to the vertical elements of the charge hall. The continuous brick plinth in a dark charcoal brick matches the bands of black sheeting above.[100]

The desire to standardise the AGRs was apparent in the architecture of the later stations, and although due to local topography no two are identical, familial similarities are evident. A common feature of the four later stations was the six crane towers used to construct and fit out the reactor buildings which remain visible as large, external buttresses. At Hartlepool and Heysham 1, the reactor buildings were simply finished in fair-faced concrete and mill-finished aluminium sheeting, which is preferable in coastal locations. To add some colour, the seaward side of their reactor halls was finished in buff-coloured panels with brown side facings (*see* Figs 6.14 and 6.17).

At Heysham 2, the architects, Powell Moya & Partners, chose to indicate the functions of the reactor building in bold horizontal layers reinforced with colour. At ground level the outer walls of the heavily shielded reactor and fuel handling building were left as exposed concrete with a textured surface and visible horizontal construction joints. Above this, on the southern elevation, the middle mechanical zone is marked by exhaust and ventilator openings (*see* Fig. 6.17). The upper-stage charge hall is clad in white profiled aluminium sheeting with an upper glazed band to provide lighting. To soften its appearance, it was given rounded corners. To the north the large turbine house was also clad in black aluminium sheeting with a wide glazed panel on its north elevation. Around the main structures, smaller buildings were finished in green and stone-coloured sheeting.[101]

Advice on the landscaping of the area in the vicinity of Heysham 2 was provided by Derek Lovejoy & Partners. The practice had also been responsible for the award-winning landscape design at West Burton (Nottinghamshire) coal-fired power station which opened in 1968. To screen its presence from the holiday camp to the south, and a golf course to the east, mounds were created and planted informally with shrubs and trees. With the docks to the immediate north and open sea to the west, it made less visual impact in those directions. Within the station fence the car park was softened with grassy areas and tree planting, although levelled areas of temporary construction camps remain to be remediated.

Extensive glazing was employed at Hunterston B (*see* Fig. 6.13) to clad the charge hall over the reactors, while the central refuelling section was indicated using metal sheet cladding. This design both reflected the form of the earlier Magnox station with its glazed boiler units and lightened an otherwise solid block against the hills to the rear. Similarly, at Torness, to break up the mass of the reactor building the sides of the charge hall were glazed and lit at night to great effect.

Howard Lobb was also the consulting architect at Torness, where, sited on a coastal headland, the visual impact was lessened by its compact form and, where possible, the use of low buildings. Its appearance was also reduced by enclosing most of the plant, which could be seen as visual clutter; this also helped to minimise the effect of salt spray on transmission lines. To further decrease the station's presence, it was faced in light grey, using translucent panels to light the reactor hall.

To protect the visual amenity of the coast, the 400kV lines were taken in underground tunnels for the first few kilometres to a point near Thornton Bridge before being carried on slimline masts. The adjacent construction camp and its site were cleared, and the land graded. In the wider landscape 1,000 trees were planted, and the South of Scotland Electricity Board worked closely with the local council to create a coastal path with viewing points facing this new landscape feature.

Sizewell B pressurised water reactor

With foresight, enough land had been acquired at Sizewell (Suffolk) for the construction of up to three stations. The delayed realisation of this potential, however, mirrored wider indecision in the industry. In November 1969 ministerial approval was granted for a second nuclear power station of the AGR type with up to four reactors.[102] The Nuclear Power Group won the contract. However, the CEGB's lower estimates for mid-1970s electricity demand suspended the construction decision in 1971, by which time some preliminary groundworks had been undertaken.[103] This pause gave the newly formed National Nuclear Corporation (NNC) the opportunity to develop the choice of station types, and in 1973 the CEGB made new applications for the use of a steam-generating heavy-water reactor (SGHWR), a high-temperature thermal reactor, or a light-water reactor system.[104] In 1974 the government announced its preference for the British SGHWR, and in the following

Fig. 6.23
Sizewell, Suffolk. A sectional 1:20 scale model of the proposed pressurised-water reactor assembled at the project board's headquarters in Knutsford, Cheshire, 30 January 1985. Models to check the position of internal fixtures were an important part of the design process and, later, important training aids to assist personnel in orientating themselves in complex plant. [National Nuclear Corporation, NCC-C458-30 Jan 85]

Fig. 6.24
Sizewell, Suffolk, 15 May 1985. A model showing construction work at month 36. This view from the east shows the hemispherical containment vessel, to its right the still roofless fuel-handling building, and to the left various plant rooms. The two chimneys are the exhaust silencers for the standby diesel generators, and to their rear is the control room. It also includes temporary features, such as cranes, to verify sequencing and access issues during construction. The large blue building is the turbine hall. [National Nuclear Corporation, NCC-C1076-15-15 May 85]

year gave its consent for this type of reactor, only for the decision to be reversed in February 1978 in favour of a PWR type.[105]

In February 1981 the CEGB made a formal application to the Secretary of State for Energy to proceed with the country's first PWR. At March 1982 prices it was estimated that the new 1,110MW station would cost about £1,147 million, against a cost of £1,590 million for a 1,250MW advanced gas-cooled reactor.[106] Sizewell offered a low surrounding local population density, good road infrastructure, an existing transmission line connection and land owned by the CEGB.

Public inquiry

The public inquiry for Sizewell's PWR was held at Snape Maltings under the chairmanship of Sir Frank Layfield. It started in January 1983 and lasted for 340 days at a cost of £20 million. The county council offered no objection to the new station, and the main issues focused on safety, economics and the environment.[107] Objectors to the new station included the Friends of the Earth, the East Anglian Alliance Against Nuclear Power, and the Stop Sizewell B Association. The Suffolk Association of Architects entered objections based on visual intrusion and the need for investment in energy conservation.[108]

The Sizewell B project received a further setback in April 1986 when the Chernobyl disaster cast further doubt on the future of the nuclear power programme. Uncertainty about potential orders again raised worries about the threat to the nuclear supply industry. During an adjournment debate in the House of Commons, Bernard Conlan, Labour MP for Gateshead East, voiced concerns for the prospects of Northern Engineering Industries' nuclear power plant division. The shortage of orders for this one company had reduced its labour force from 1,200 in 1984 to just 137 in 1986.[109] In July 1986, prior to the inquiry's decision, the CEGB announced that if the Sizewell B was successful, they planned to build a further five or six PWRs based on the Sizewell design.[110] Other locations under consideration for PWRs included Winfrith (Dorset), Druridge Bay (Northumberland), and adjacent to existing stations at Hinkley Point and Dungeness.[111]

The final report was published on 26 January 1987; arguments had ranged around future predictions for energy consumption and reflecting on alternative energy sources. Although some concerns were raised about safety, the economic case was accepted. It was argued that PWR technology would be 25 per cent cheaper than the AGR design and was the preferred option of the CEGB.[112] The examination also represented a major public discussion about the future of nuclear power, and effectively marked the end of advanced gas-cooled reactor technology.

Politically, lessening the dominance of coal for power generation and the vulnerability of its supply chain to industrial action by the miners and transport workers was also attractive.[113]

Design

Design work for the new PWR station was carried out by NCC, making full use of lessons learnt in the United States but following United Kingdom safety requirements and engineering practice (Figs 6.23 and 6.24). The CEGB specified that an architect–engineer familiar with the Westinghouse system should be retained by NCC; this service was provided by the American company Bechtel. The design drew on the Westinghouse Standardised Nuclear Unit Power Plant Systems (SNUPPS), which as the name implies, had been producing standard designs for new US reactors. Similar stations were being built at Calloway (Missouri) and Wolf Creek (Kansas), although their layouts differed according to local site configurations, and they used fresh water for cooling instead of potentially corrosive seawater as at Sizewell.[114]

The consulting architects for the project were Yorke, Rosenberg, Mardall, led by Brian Henderson. Speaking later, he said his advice on power stations was to 'keep the design simple'.[115] This he held to at Sizewell, with the unassuming blue and white metal-clad buildings and the low-maintenance ceramic tile-covered dome. Peter Youngman, who had been responsible for the landscape design of Sizewell A, was retained to advise on the new station. It was conceded that little could be done to hide the structures, but for the first time the CEGB was willing to consider the wider offsite landscape and, using agreements and grants, woodland, hedgerows, tree belts and roadside trees were planted to screen settlements and to enhance views from well-trodden paths.[116] Planting within the station's boundary fence was used to screen car parks and ancillary features.

Construction

In 1984, the CEGB announced that it would not start work on the foundations until spring 1986, and further construction work on the station would not begin until September that year. This was in part due to delays caused by the inquiry, the necessity of acquiring specialist parts, and the time taken to obtain the necessary safety clearances. However, in February 1984, before the inquiry had reported, the CEGB was given clearance to start ordering major components, such as the steel reactor vessel from Framatome, France. The Board argued that to delay would be more costly than to go ahead, with the risk that the scheme would be cancelled afterwards.[117] The employment benefits of constructing the station were spread across the United Kingdom. It was estimated that the supply of components would employ about 10,000 people for seven years, while the peak workforce on site would number about 3,000 people contributing 24 million hours of work.[118]

Station layout

The station and the local landscape are dominated by the large white hemisphere of the reactor containment structure (Fig. 6.25). Conjoined to this is a group of anonymous buildings clad in blue and white sheeting that house the station's key components, with only the differences in roof lines indicating their functions (Fig. 6.26). To the south of the reactor is the long mechanical annexe, and adjacent to it the large turbine hall. This houses two GEC turbine generator sets, each rated at 660MW, of a similar design to those used at the AGR and fossil fuel stations. On the seaward side of the reactor are the auxiliary building, control room and the emergency generators, indicated by their exhaust stacks, and straddling the site boundary is the cooling-water pumphouse. Attached to the north of the reactor dome is the fuel building, used to handle and store new and irradiated fuel, and the long radioactive waste processing and storage building. Surrounding the central complex are white-clad, single-, two- and three-storey support buildings including, administration, workshops and welfare. To the west, due to its coastal location and the risk of salt deposition causing electrical shorting, the 400kV substation is enclosed in a separate structure.

The reactor

Sizewell B is based on a standard Westinghouse design with four reactor coolant circuits or loops. The water is pressurised to 150 bar, and each loop contains a steam generator and pump. The reactor vessel is constructed from forged and welded construction and is clad internally in corrosion-resistant steel. This sits in a vault in the floor enclosed by the cylindrical reinforced concrete containment vessel with a hemispherical roof, which also encloses the nuclear steam supply system comprising four steam generators. To prevent any small amounts of leakage, the concrete containment vessel is lined in steel

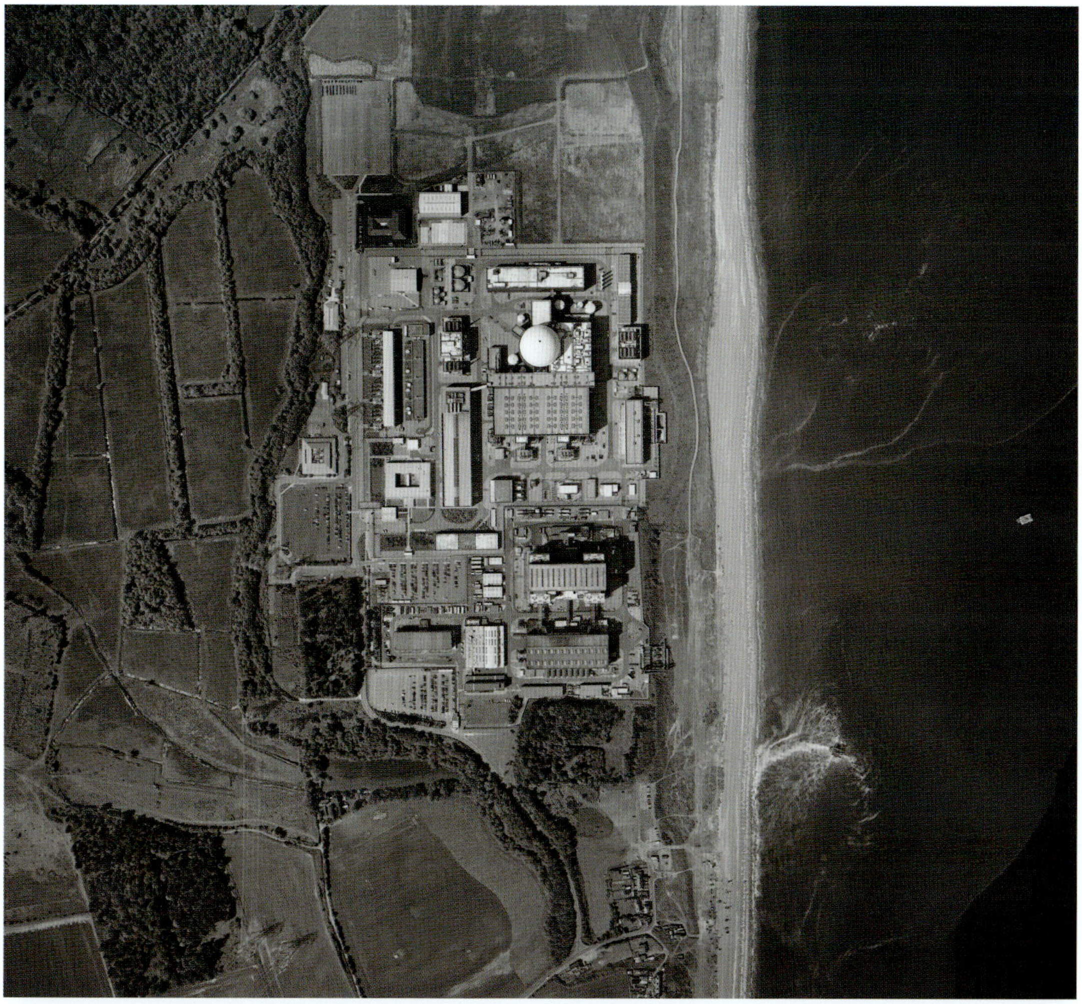

Fig. 6.25
Sizewell, Suffolk. To the top is the dome of Sizewell B PWR and to the bottom the Magnox station. Woodland was planted around the station on the landward side to screen the car parks and smaller buildings.
[© Crown copyright. Ordnance Survey os_99143_V_057]

Fig. 6.26
Sizewell B, Suffolk. The view is dominated by the white ceramic tile-clad hemispherical containment structure which houses the PWR.
[© Historic England Archive, 26275_013]

Fig. 6.27
Sizewell, Suffolk, August 1992. View showing the hemispherical containment structure and the beginning of the assembly of the outer steel containment shell.
[© Historic England Archive (John Laing Photographic Collection), JLP01_11_54331_12]

plate and further protected by a secondary outer steel containment shell covered by white ceramic tiles (Fig. 6.27). Internally, at the top of the cylinder is a Polar crane that rotates on a circular monorail which was used for construction and remains available for maintenance tasks. The plant is protected by four independent safety systems, twice as many as in the original SNUPPS design. They comprise four high-pressure safety injection pumps, one coupled to each loop, four accumulators containing water under pressure, four residual heat removal pumps, and two heat exchangers. There are also four emergency generators and an auxiliary shutdown room separated from the main control room.[119]

The fuel in the reactor comprises cylindrical pellets of enriched uranium oxide contained in standard 9.5mm diameter, 3.85m long Westinghouse fuel rods clad in zirconium alloy; 264 of these rods are supported in a stainless steel fuel assembly, 193 of which are loaded into the core.[120] Reactivity is managed by control rod clusters and by adding neutron-absorbing boric acid to the reactor coolant water. On arrival on site, fuel assemblies are taken to the fuel building on the north side of the reactor building. Fuel assemblies typically spend three years in the reactor. Each year one-third are replaced by new assemblies, and the remainder rearranged to ensure the most efficient fuel usage. Refuelling is undertaken annually, usually in the summer months, and involves removing the top of the reactor vessel. Irradiated fuel assemblies are then removed by the refuelling machine and placed in borated water within the refuelling cavity and then moved on through other storage ponds before transfer to transport flasks.[121] At the north end of the site, oriented east to west, is the radioactive waste process and the relatively large storage building designed to hold spent reactor fuel.

The control room is housed in a separate building to the east of the reactor. This is generally staffed by an operator and supervisor who control the reactor and the turbines. Other rooms in the building contain data loggers and reactor safety systems. In common with other complex technologies, simulators at the Oldbury-on-Severn Nuclear Training Centre and on site at Sizewell are used for personnel training.

Conclusion

In total, 15 advanced gas-cooled reactors (AGRs), including the Windscale pilot plant, were built in the United Kingdom. Although they represented a technological development from Britain's early research reactors and the Magnox series, the programme was marred by problems from the start. Technical issues, labour disputes among construction workers, and lower than anticipated demand for electricity, compounded by a financial background of rapidly rising inflation, all culminated in commissioning delays and exorbitant cost overruns. By the early decades of the 21st century, Britain was the only user of this technology, which represented only around 3 per cent of the global reactor fleet. In the 1980s, additional PWR stations were planned for Hinkley Point, Sizewell and Wylfa, but with the privatisation of the electricity industry these schemes were halted. Internationally, the pressurised water reactor technology of Sizewell B represents the most common type of reactor system, and this will be used at the new Hinkley Point C and most likely for any further new nuclear stations.

Box 6.1 Industry as spectacle

Since the beginnings of the industrial age, travellers, such as Celia Fiennes in the 17th century, have been intrigued by the mysteries of industrial processes and their landscapes, which in the following century drew painters to the fiery furnaces of Ironbridge Gorge (Shropshire). In Victorian Britain, huge crowds flocked to see the new technological marvels, from the opening of the Liverpool to Manchester railway to the launch of Brunel's great iron ship the *Great Eastern*.

Gabrielle Hecht has discussed how the French nuclear industry might be seen as a technological spectacle, a drama of national salvation with sites presented as tourist attractions to be visited as one might a museum or ancient monument.[1] As the British nuclear power programme emerged from the secrecy surrounding its early associations with atomic weapons, there was great pride in what was being achieved and the benefits the technology might bring. The United Kingdom Atomic Energy Authority (UKAEA) was eager to communicate this excitement to the public and to maintain political support for its programmes. In London, in the late 1950s, UKAEA built a prestigious eight-storey headquarters close to the heart of government off Regent Street (Fig. B6.1.1). With unashamedly modern interiors by the renowned industrial designer Kenneth Grange, on the ground floor was a small exhibition room dedicated to explaining the marvels of nuclear science to interested passers-by.[2]

Such was the novelty and scale of construction of the new stations that they attracted enthusiastic attention from the professional technical press. The general public was also fascinated by these modern industrial wonders; at Berkeley power station (Gloucestershire), 20,000 people visited the construction site between 1957 and 1962.[3] Once in operation, stations continued to excite interest, and guided tours became an important way for the industry to debunk fears and promote its message. The opening of the new atomic stations also coincided with the growth in car ownership, making relatively remote stations accessible to many people.

UKAEA's facilities also provided an important technical education role, especially following the announcement of the first civil nuclear power programme, through which the government wished to promote a strong private-sector industry with nuclear expertise. On one

Fig. B6.1.1 United Kingdom Atomic Energy Authority Headquarters, Charles II Street, London. Design by Leslie C. Norton, Trehearne & Norman, Preston & Partners. Trollope and Colls, one of the main contractors at Windscale, were responsible for the building work. [TNA BP2/72. Contains public sector information licensed under the Open Government Licence v3.0]

Fig. B6.1.2
Dounreay, Caithness, May 1954. Mothballed former Royal Naval Air Service control tower. It was later used as the headquarters for the construction project before being converted into a visitor centre. Storm damage led to its demolition in 2007.
[© Historic England Archive, G5163_1]

Fig. B6.1.3
Sellafield Visitor Centre, Cumbria, 2006.
[© Wayne Cocroft]

occasion, in July 1957, the north-west branch of the Institution of Mechanical Engineers arranged a visit to Calder Hall for 580 of its members and 'their ladies'. Most travelled by train to Sellafield, although it was beyond the resources of UKAEA to provide tea for so many.[4] The author recalls a similar visit by an English Heritage party in the late 1990s when it was still possible to stand on the charge face with the reactor running beneath.

In June 1960 the *John O'Groats Journal* commented that the Dounreay Experimental Research Establishment was 'a prize any tourist-minded county would jump at'. UKAEA was quick to recognise the educational benefits of converting the airfield control tower into a visitor centre, while keeping the curious at least 400m from the perimeter fence (Fig. B6.1.2). During the first tourist season 8,000 people visited.[5] By the 1970s and 1980s, visitor centres became a common feature at many nuclear power stations (Figs B6.1.3 and B6.1.4). Generally, they were placed adjacent to a station's main car park and outside its security fence. They presented the casual visitor and school parties with lively graphical displays, models, objects and films. For more determined visitors it was usually possible to book escorted tours behind the fence, in some cases changing into suitable clothing and footwear before gaining access to the reactor hall.

The visitor experience became part of the architecture of the stations. At Trawsfynydd (Gwynedd), adjacent to the lake a car park, fishing club and cafe were provided, along with a viewing platform. Inside, a visitors' viewing room was placed above the charge hall. These also became a feature of the later stations at Heysham (Lancashire), Hartlepool and Torness (East Lothian). At Heysham, the local architects Harrison Pitt Architects were commissioned to design a high-profile visitor centre as the public face of the station. Opened on 1 August 1995, the result was a steel-framed building clad in dark- and pale-blue profiled aluminium sheeting, surrounded by car parks and landscaping. A light and fully glazed foyer accommodated the reception area, shop and cafe. The rear was windowless and housed an exhibition area, while the first floor housed offices, conference rooms and training facilities.[6] A nature park was created to the east of the complex, primarily to benefit local schools, with an observation tower overlooking the two stations. This openness was brought to a sudden end in September 2001 when the terrorist attacks on the World Trade Center in New York led to a major re-evaluation of nuclear security. This event also coincided with the decommissioning of the Magnox stations, and later the COVID-19 pandemic of 2020 led to further closures. Today EDF welcomes people to its visitor centres at its principal operational power stations, with exhibitions aimed at school groups.

Fig. B6.1.4
Battle of the badges. Souvenirs from the Sellafield Visitor Centre from the turn of the 21st century; each measures about 55mm in diameter. [Author's collection]

7

Postscript

The Nuclear Decommissioning Authority (NDA) was established by the Energy Act 2004 as a non-departmental public body sponsored and funded by the Department for Energy, Security and Net Zero (DESNZ). It owns the United Kingdom's 17 earliest civil nuclear sites with a mission to clean up the sites safely, securely and cost-effectively, leaving them ready for their next use. The initial remediation tasks will occupy the next century, with an end state for all current sites predicted to be 2333. Over the next decade or so, it will also become responsible for the seven advanced gas-cooled reactor stations.[1]

In addition to the sites and their sometimes hazardous waste, the NDA also acquired responsibility for a vast collection of plans, drawings, photographs, film, microfiche and documents, representing some of the earliest archive material from the birth of the atomic age. Usually, these public records would pass to The National Archives, but by agreement they are being deposited instead at Nucleus: The Nuclear and Caithness Archives, Wick (Fig. 7.1).[2] Here records 'lifted and sifted' from the NDA's sites across the country are being catalogued and eventually made available to researchers.

Dounreay was one of the earliest sites to enter decommissioning, which quickly brought questions about the future of the fast reactor dome. Although a rare building form, it had become an instantly recognisable symbol of the industry. To address its future and the historical legacy of the whole site, a heritage strategy was commissioned.[3] This recognised the significance of the site's structures but concluded, due to contamination issues, long-term maintenance and security issues, that retention was not an option. Instead, during demolition, their historic value will be captured through a programme of building recording. The strategy also recognised the significance of fixed and portable objects, digital and paper records, and the intangible memories of its staff. The importance of sharing and communicating this heritage was also highlighted. Locally, the story of Dounreay is told at the North Coast Visitor Centre, Thurso, which displays the control desk from the Dounreay Material Test Reactor. The control desk from the Dounreay Fast Reactor has been acquired jointly by National Museums Scotland.

In the United Kingdom, ultimately it is likely that few, if any, historic civil nuclear facilities will remain. The power stations will be dismantled, their non-nuclear components quickly demolished and cleared. Only the reactor buildings, containing the most heavily irradiated

Fig. 7.1
Nucleus: The Nuclear and Caithness Archives, Wick.
[© Wayne Cocroft]

components, will remain, partly stripped, lowered and reclad to maintain their security until final demolition decades in the future (Fig. 7.2). Many power stations sat within designed landscapes, which, like earlier garden landscapes with now-lost country houses, will endure, but without the focal point of the tall reactor structures. On the research sites, many buildings were too specialised or contaminated for reuse, but their locations have been reimagined as science and technology centres (*see* Fig. 4.13). A handful of laboratories associated with the understanding of the atom are protected for their historic interest, often due to their intrinsic architectural merit. Some of the places connected with early research into the atomic bomb are also protected, as are several sites associated with the deployment of nuclear weapons: information on these may be found on the National Heritage List for England.

Other physical legacies include irradiated waste produced by reprocessing fuel assemblies, and plant contaminated through contact with radioactive substances. Many operations involving radioactive sources will also generate low-level waste. These include healthcare, materials research, industrial quality control, and clothing worn by workers. This type of waste is held by a single national low-level waste repository at Drigg (Cumbria) about 5.5km south-east of Sellafield (Fig. 7.3). Originally, the waste was tipped into open trenches, which have subsequently been capped with stone and earth. In recent decades, this method has been replaced by a system using standard and stackable containers. These arrive by road or rail, and are taken to the cementation plant where any remaining voids in the containers are filled with a cement grout. After this has hardened, the containers are moved to engineered concrete vaults where, depending on the depth of the container, they may be stacked at least six high. Ultimately, the intention is to cap these with earth and rock, in so doing creating a new landform, beneath which will lie an entombed and inaccessible archaeological assemblage of Britain's atomic age.

The long-term issue of the safe storage of nuclear waste, including the country's stock of plutonium, is yet to be resolved. The international consensus is that deep geological storage offers the best solution. The challenge is to find rock formations that will be stable potentially for hundreds of thousands of years, communities willing to accept the associated above-ground facilities, and the creation of enduring signals to alert future generations to the hazards that lie within.

If the survival of structures is unlikely, is another option to select objects for museum displays to tell the story of the atomic age? Collecting objects to represent modern industries presents many challenges. Curators need to be aware of the location of items, their technological

Fig. 7.2
Bradwell, Essex, 2017. Reduced and reclad reactor buildings; compare with Fig. 5.10.
[© Wayne Cocroft]

Fig. 7.3
Drigg, Cumbria, 2011. In the foreground is the low-level waste repository, showing the rectilinear layout of the wartime TNT factory, and, above, an engineered vault. In the middle distance is Seascale, and Sellafield beyond. [© Historic England Archive, 28127_020]

Fig. 7.4
Orford Ness, Suffolk. WE177A tactical freefall nuclear weapon displayed by the National Trust. [© Historic England Archive, DP068440]

significance and when they might become available. If they are obtainable, they may be heavy, large, contaminated and difficult to remove; any activity connected with the nuclear industry is likely to attract greater costs.

In the United Kingdom, the Science Museum, London, and the National Museum of Scotland, Edinburgh, have included atomic energy in general exhibitions, and local museums with nuclear sites close by, such as the Beacon Museum, Whitehaven, may have small displays. The main collection of artefacts relating to the development of the country's nuclear deterrent remains restricted within the Atomic Weapons Establishment, although nuclear weapons casings, in a few cases with exposed warheads, are displayed in several aviation and military museums. Orford Ness, owned by the National Trust, is a rare example of a freely accessible nuclear test site (Fig. 7.4).

International nuclear heritage

Internationally, there is a mixed picture of how nuclear heritage, and especially its built infrastructure, is being preserved and presented: space here only permits a very brief survey. Common responses include the presentation of nuclear technology, usually civil, in national science museums. At a local level there are often institutional initiatives to preserve reminders of important innovations and discoveries. However, it is also a heritage that can remain restricted, either to workers within the industry, or at least to the citizens of a given country. In France, the stainless steel dome of A1 Chinon has been used to house the Musée de l'Atome, a visitor centre for the operational nuclear plant.

Fig. 7.5
Project 816, Fuling, Chongqing Province, China. Entrance to underground plutonium production facility.
[© Wayne Cocroft]

North America has the most extensive assortment of accessible nuclear heritage sites, and multiple examples of nuclear weapon casings and missiles are on display in aviation museums. The most ambitious designation is the Manhattan Project National Historical Park, a joint venture between the National Parks Service and the US Department of Energy.[4] Spread across three states, the park encompasses some of the most significant components of the wartime Manhattan Project. This includes the Hanford B reactor, (Washington); the world's first full-scale reactor, the X-10 plutonium reactor, Oak Ridge (Tennessee); and specialised research and test structures at Los Alamos (New Mexico). Beyond the key locations, other facilities are often preserved within these large government enclaves, and local initiatives highlight other features through tourist trails or more specialised museums. In Los Alamos, for example, the Bradbury Science Museum explores the work of the Los Alamos National Laboratory, while the local historical society presents the buildings of the ranch school used by the project.

The most comprehensive museum dedicated to nuclear history is the National Museum of Nuclear Science, Albuquerque (New Mexico). Opened in 1969, it had its origins in an institutional museum at Kirtland Air Force Base, before security concerns forced its move to an off-base location.[5] Although its emphasis remains on nuclear weaponry, its exhibitions also present nuclear history, power, medicine, uranium mining, energy and radiation. In Nevada, tours are offered of the National Security Site, formerly the Nevada Test Site, and, more accessibly, just off Las Vegas Strip is the National Atomic Testing Museum. The United States' contribution to civil nuclear power is commemorated by the fast breeder EBR-1 (Idaho), which first generated power in December 1951. This was designated as a National Historical Landmark in 1966 and is open for tours.[6]

Russia too has a wide range of nuclear heritage sites and museums, with Rosatom, the State Atomic Energy Authority, overseeing over 100 museums. The majority of these are small exhibitions set up by the staff of individual establishments and have parallels with tradition rooms in other socialist industries and armed forces. In Moscow, other exhibitions from the 1950s at the Polytechnical Museum and Exhibition of Achievements of National Economy had parallels in the west with an emphasis on the peaceful uses of nuclear technology. Also in Moscow, the first Soviet reactor to achieve criticality on 25 December 1946, F-1 in the Kurchatov Institute, is preserved and may be visited. Just over 100km away in Obninsk, a small reactor connected to a local power grid that operated between 1954 and 2002 has been preserved and made accessible, as has the decommissioned atomic-powered icebreaker, *Lenin*, in Murmansk. In line with the central message that the Soviet world was only interested in the peaceful exploitation of the atom, no nuclear weapons were displayed until the early 1990s.[7] Prior to 2022, the site of the world's most notorious nuclear accident at Chernobyl, Ukraine, was developing as a niche tourist destination, with a dedicated museum in Kyiv.

In China, sections of the 20km tunnels excavated to house a plutonium production reactor and associated processing plant, Project 816, Fuling, Chongqing Province, have been opened as a tourist attraction (Fig. 7.5). The motivation behind the enterprise is to create a tourist infrastructure in addition to strengthening national, military and scientific education, and celebrating the achievements of the construction workers. Similarly, the Red Mountain Command Base of the test range on the edge of the Taklamakan desert, Xinjiang Province, has been opened for tourists.[8]

Nuclear futures

It is over 70 years since the United Kingdom detonated her first atomic weapon; the principal responsibilities of the Atomic Weapons Establishment today are to maintain the surety of the current Mk 4/A (Holbrook) warhead and the development of a potential successor. To support these programmes, major technical infrastructure campaigns continue, including at Aldermaston, where Project Pegasus is to provide a new enriched uranium store and manufacturing facility with an approved cost of £634 million, and close by at Burghfield, Project MENSA, a new warhead assembly and disassembly plant with a projected cost of £1,996 million.[9]

As part of the government policy to achieve carbon net zero by the middle of the century, the

Fig. 7.6
Hinkley Point, Somerset, August 2020. Work underway on Hinkley Point C. To the top left are the decommissioned A and B stations.
[© Historic England Archive, 33900_003]

United Kingdom is now committed to a programme of new nuclear power stations.[10] Work is underway on a pair of European pressurised water reactors at Hinkley Point (Somerset) with a projected power output of 3.2GW (Fig. 7.6). A further location has been identified at Sizewell (Suffolk) and a third site will be sought. Given the challenging combination of geographic factors that influence the siting of nuclear facilities, decommissioned sites may be reassessed for building new reactors. Small modular reactors may also permit the construction of stations closer to demand centres in locations not previously associated with the industry. To secure future fuel supplies, it's likely that facilities at Capenhurst (Cheshire) and Springfields (Lancashire) will be updated, although there are no current plans to revive reprocessing. Historically, beyond the power stations and research centres, were webs of supply chains stretching across the country, and these are being renewed.

Notes

Chapter 1

1 Aylen 2015, 31–59.
2 Cocroft and Alexander 2009; Cocroft and Newsome 2009; Cocroft 2010.
3 Bone et al. 2006; Litherland 2006; Nichol and Litherland 2006.
4 TNA: POWE 14/1244, Minutes of Public Inquiry, 54.
5 www.nhm.ac.uk/discover/what-is-the-anthropocene.html (acc. 23 November 2023).
6 Davies 2012.

Chapter 2

1 Andrade 1964, 99, 125–6.
2 Nuttall 2005, 30.
3 Oliphant and Penney 1968, 140.
4 Cathcart 2004, 259.
5 Ibid., 201–7.
6 Andrade 1964, 194.
7 https://collection.sciencemuseumgroup.org.uk/objects/co5130/cockroft-and-waltons-accelerator-particle-accelerator (acc. 12 April 2024).
8 Clark 1980, 28–9.
9 Oliphant and Penney 1968, 146–7.
10 www.cambridgephysics.org/laboratory/laboratory13_1.htm (acc. 22 November 2023).
11 Clark 1980, 43–4.
12 Gowing 1987, 8–9.
13 Clark 1961, 35–7.
14 www.liverpoolmuseums.org.uk/artifact/cyclotron-dees (acc. 2 February 2024).
15 Clark 1961, 44, 46, 47, 84, 116.
16 Ibid., 42, 49.
17 Ibid., 49–56.
18 Memorandum reproduced by Gowing 1964, 389–93; Reader 1975, 287.
19 Smyth 1945, 31; Gowing 1987, 14; The meaning and rendition of the codename MAUD has been open to several interpretations. It has been written as Maud, maud, M.A.U.D. and MAUD. Some claim it was derived from the name of Neil Bohr's English nanny Maud Ray. Or, it might be an acronym for Ministry of Air Uranium Development or Detonation, or Military Application of Uranium Detonation. For MAUD report *see* Gowing 1964, 394–438.
20 Clark 1961, 89–91.
21 Bone et al. 2006, 54.
22 Clark 1961, 117.
23 Reader 1975, 189–90.
24 Clark 1961, 69–73, 95–103; Gowing 1987, 17.
25 Reader 1975, 287–93.
26 Moore 2021, S376.
27 Clark 1961, 155.
28 Identified in letter TNA: AB 1/68 2/2/2 31 Dec 1941.
29 Clark 1961, 134.
30 Reader 1975, 289.
31 Clark 1961, 129, 174.
32 Bone et al. 2006, 55–6.
33 TNA: AB 1/68 drwgs 10545/5 8.1.42, 10545/10 21.2.42.
34 Clark 1961, 175.
35 TNA: AB 1/68, 2/2/2 25 Feb 42.
36 Litherland 2006 drwg RHYD 98004.
37 Litherland 2006, 9–13.
38 Bone et al. 2006, 57.
39 Ibid., 57–8; MAUD Report 1941, 13, https://fissilematerials.org/library/maud.pdf describes a rotor diameter of 72cm (acc. 16 April 2024).
40 www.nps.gov/places/000/k-25-gaseous-diffusion-plant.htm (acc. 22 November 2023).
41 Clark 1961, 176.
42 Clark 1980, 119.
43 Hughes 2002, 97; Neufeld 1995, 273.
44 Reader 1975, 293.
45 Smyth 1945, 30–1.
46 Gowing 1987, 16.
47 *See* Gowing 1964, 439–40.
48 Oliphant and Penney 1968, 162.
49 Clark 1980, 130.
50 Gibson 1996, 86.
51 Smyth 1945, 31.

Box 2.1

1 NRPB 1989, 7.
2 Martin and Harbison 1986, 131–2, 135–6.
3 Ibid., 133.
4 Radford 1961, 75.
5 Buckley 2004, 4.

Chapter 3

1 TNA: AIR 2/13689.
2 TNA: AB 8/557, 1–2; Cathcart 1994, 23–4.
3 Jay 1954, 6.
4 Ibid., 1.
5 Oliphant and Penney 1968, 164.
6 *Electrical Review*, 1 July 1955, 29–30; *Electrical Times*, 30 June 1955, 1061; Francis and Crisp 2008, 167.
7 Jay1954, 7.
8 For a fuller account of the uranium production process, *see* Jay 1954, 11–20.
9 Arnold 2007, 8–9.
10 TNA: AB 8/557.
11 Jay 1954, 22; Arnold 2007, 10.
12 TNA: AB 6/326 Draft minute 11/1/1/46 2 May 1947.
13 Ibid.; TNA: BP 2/145 letter 12 January 1948.
14 TNA: AB 8/557; Arnold 2007, 11.
15 Viney 1951, 3.
16 Dick 1954, 303.
17 TNA: AB 58/8 Report on Calder Bridge 9 August 1947; Jay 1954, 23.
18 Viney 1951, 5.
19 Dick 1954, 303.
20 Arnold 2007, 15, 17–18.
21 Dick 1954, 289.
22 Viney 1951, 8.
23 Dick 1954, 290.
24 Arnold 2007, 16.
25 Ibid., 15.
26 Viney 1951, 10.
27 Gill 1964, 28–35.
28 Hall 1963, 30.
29 Arnold 2007, 31–2, 168.
30 Ibid., 40–52.
31 Ibid., 66–8.
32 Ibid., 86.

33 Ibid., 124.
34 Ibid., 133.
35 Ibid., 34, 40, 154–5.
36 Ibid., 73–4, 117, 121, 123.
37 Penney 1968, 297.
38 Dick 1955, 528.
39 Jay 1954, 34.
40 Jay 1955, 5.
41 Jay 1954, 32–41.
42 Ibid., 38.
43 Hansard, HC Deb., vol. 477, col. 115 (19 June 1950); Hansard HC Deb., vol. 498, col. 212 (8 April 1952).
44 Jay 1954, 54.
45 Ibid., 50.
46 Hansard, HC Deb., vol. 722, cols 624–8 (9 December 1965).
47 *Electrical Review*, 23 September 1977, 5; Norris et al. 1994, 78–9.
48 Cocroft 2010.
49 Gowing 1974, 442.
50 TNA: ES 1/272 E13 High Explosives Research 10 April 1947.
51 Gowing 1974, 130, 452.
52 Griffiths 1984, 2.
53 DSTL Fort Halstead F16 AB 1/1 5839/21A Bomb Chamber 20 July 1947; F16 H E Research Bomb Chamber Lightning Protection Installation 12.9.47.
54 DSTL Fort Halstead F17 AB 2/1 5839/32A Fort Halstead H E Research Detonation Laboratory August 1947.
55 Hughes n.d., 25–6.
56 Colin Hughes, pers. comm.
57 Wynn 1994, 29–31.
58 Hughes n.d., 28–9.
59 Rowlands 1990, 14.
60 Wynn 1994, 39.
61 Rowlands 1990, 15.
62 TNA: ES 1/246, establishment table 1.12.50.
63 Ibid.
64 TNA: ES 1/272 E7.
65 TNA: ES 1/272 E10A 6 Oct 1954.
66 TNA: ES 1/1420 letter 24 1 55; TNA: ES 1/691 letter 31 5 1956; TNA: AB 16/1890 letter 7/12/1956; Newsome and Williams 2011.
67 TNA: ES 1/691 'Expansion of AWRE Woolwich Common' note by William Penney 20/1/56; TNA: AB 46/10 Letter 16 May 1963; TNA: AB 44/6 Letter 13/9/1963.
68 TNA: ES 1/1420 letter 24 1 55; TNA: ES 1/691 letter 31 5 1956; TNA: AB 16/1890 letter 7/12/1956.
69 TNA: ES 1/691 letter 20/1/1956.
70 TNA: ES 1/272 Memorandum 10 October 1952.
71 TNA: ES 1/246 table 1 12 50 ARE strength as at 1 12 50; TNA: ES 1/691 letter 20/1/56, TNA: ES 1/691 'Woolwich Common extension: buildings and works services', note by William Penney 20 1 1956.
72 GLC/AR/BR/26/WO/M/241-2.
73 Ibid
74 TNA: AB 46/10 letter 16/5/1963; TNA: AB 44/6 letter 13/9/1963.
75 Norris et al. 1994, 72.
76 Ibid., 72.
77 *Electrical Review*, 14 August 1959, 17–19.
78 *Nuclear Energy Engineer*, 13, issue 130 (March 1959), 133–5; Norris et al. 1994, 70.
79 Norris et al. 1994, 70.
80 www.osti.gov/opennet/manhattan-project-history/Science/ParticleAccelerators/cockroft.html (acc. 16 April 2024).
81 *BEAMA*, August 1959, 87; *Electrical Review*, 14 August 1959, 17–19.
82 Lavington n.d., 4.
83 Taylor 2001, 52.
84 www.ibm.com/history/stretch (acc. 23 December 2023).
85 Taylor 2001, 53.
86 Ibid., 53–4.
87 Lavington n.d., 7.
88 Ibid., 4.
89 Coolidge 1996, 15.
90 Cathcart 1994, 71.
91 Cocroft and Newsome 2009.
92 TNA: ES 1/330.
93 TNA: ES 1/267.
94 TNA: ES 1/330 Drawing 18 December 1947.
95 www.nps.gov/places/000/v-site.htm (acc. 20 December 2023).
96 TNA: ES 1/330 Memo 29 Dec 1947 PR/2/1.
97 TNA: ES 1/330; Cathcart 1994, 119.
98 TNA: AB 16/916.
99 Cathcart 1994, 182–3, TNA: ES 1/331; Pyne 2006; Bob Crump, pers. comm.
100 Pyne 2006.
101 Cathcart 1994, 175–6.
102 Norris et al. 1994, 25–6.
103 Hennessey 2003, 61; TNA: AB 16/916.
104 TNA: ES 1/269.
105 Mills 1951, 60–5.
106 Elwall 2000, 30–40; Harwood 2010, 72–5, 90.
107 Arnold 2001, 108–27.
108 Ibid., 191.
109 www.subbrit.org.uk/features/operation-orpheus/ (acc. 20 March 2024).
110 TNA: AVIA 65/1258.
111 TNA: ES 1/269.
112 Penney, W. G. letter 23 March 1955, source Aldermaston file 0239 1a. Copy held by the National Trust.
113 TNA: AB 16/2975 E4 6 Oct 59.
114 Cocroft and Thomas 2003, 30.
115 AWRE n.d., 10.
116 TNA: AB 16/2228 E45 15/1/60.
117 Millington 1971.
118 Ibid.
119 Ibid.
120 Cocroft and Thomas 2003, 244–5.
121 TNA: AB 16/2975 E139 12 June 1962.
122 AWRE n.d., 6–7, 9.
123 UKAEA 1962 and 1963.
124 AWRE n.d., 13.
125 Hennessy 2007, 124.
126 Kinsey 1981, 115.
127 Aylen 2015.
128 www.nuclear-weapons.info/vw.htm#Violet%20Club.
129 Gowing 1974, 179.
130 Aylen 2015, 40, 56.
131 W. L. Day commenting on a lecture by D. R. R. Dick, Dick 1955, 539.
132 Dick 1954, 303.
133 Henderson 1956, 509, 513.

Box 3.1

1 Hance 2006, 152.
2 Ibid., 57.
3 'Abingdon and surrounding areas facts and history', www.facebook.com/Abingdonandsurroundingareas/posts/1868360003336302/ (acc. 9 November 2023).
4 Hansard, HC Deb., vol. 508, col. 138 (2 December 1952).
5 Garrett 2016, 45.
6 Ibid.
7 Ibid., 29, 41.
8 Ibid., 40, 42.
9 TNA: AB 8/134 piece 66/75/23 24.11.47.
10 Davey 1994.
11 TNA: AB 8/272 Housing at Mirehouses Whitehaven C. A. Bird 2.1.59.
12 https://historicengland.org.uk/listing/the-list/list-entry/1473399 (acc. 14 November 2023).
13 Garrett 2016, 34–5.
14 Fry 2015, 16–17.

15 www.highlifehighland.com/nucleus-nuclear-caithness-archives/atomic-housing/ (acc. 15 November 2023).
16 Ross 2019, 188–93.
17 Laird 1957, 847; *Architects' Journal*, 11 September 1958, 391–4.
18 Bruce 2002, 5.
19 Fry 2015, 38.
20 Hance 2006, 211–15.

Box 3.2

1 Fry 2015, 10–11.
2 *Daily Mail* 1954 'Atomics in the service of mankind', 3.
3 Hennessy 2003, 107–16.
4 https://cnduk.org/peoples-history-of-cnd-easter-marches-to-aldermaston-1958-60/; www.peacemuseum.org.uk/easter-marches-1958-and-1965/ (acc. 30 January 2024).
5 Carroll 2010, 7.
6 Welsh 2001, 4.2.
7 *Electrical Review*, 12 May 1978, 11; *Electrical Review*, 1 June 1978, 16–17.
8 Smith 1986, 11.
9 https://researchbriefings.files.parliament.uk/documents/POST-PN-106/POST-PN-106.pdf (acc. 5 February 2024).
10 Welsh 2001, 12.1–12.4.

Chapter 4

1 Vig 1968, 20, 57.
2 Hance 2006, 39.
3 Oliphant and Penney 1968, 163.
4 Hansard, HC Deb., vol. 415, cols 38–9 (29 October 1945).
5 Oliphant and Penney 1968, 142.
6 Hance 2006, 204.
7 Bowyer 1983, 164.
8 Penney 1968, 297; Echo 1996, 4.
9 Penney 1968, 297.
10 Jay 1955, 71.
11 Ibid., 112.
12 *Nuclear Engineering International*, 2 August 2005 'Decontamination and Decommissioning Harwell to End Well', www.neimagazine.com/features/featureharwell-to-end-well-721/ (acc. 24 July 2023).
13 Ministry of Works 1949.
14 www.rsc.org/news-events/articles/2009/11-november/harwell-honoured/ (acc. 24 July 2023).
15 Penney 1968, 298; Hance 2006, 108.
16 Jay 1955, 60–3.
17 *The Electrician*, 30 July 1948, 316.
18 Jay 1955, 112–13; Hance 2006, 276.
19 Hance 2006, 113.
20 *Combustion, Boiler House and Nuclear Review*, February/March 1958, 76–7; *Atomics & Nuclear Energy*, 1958, 9, issue 2, 60–1; *Nuclear Engineering*, 1958, 3, issue 2, 121–2; *Industrial Architecture*, 1958, 1, issue 1, 58–60.
21 Penney 1968, 299.
22 *Nature*, 11 December 1954, 1085.
23 *Electrical Times*, 14 December 1950, 946–7; Hill 2013, 199.
24 During the war faltering attempts had been made to build reactors in Germany at Kummersdorf and Haigerloch, Baden Württemberg; Hance 2006, 65.
25 Hill 2013, 202–3.
26 Echo 1996, 6.
27 Hance 2006, 47.
28 Hill 2013, 36, 204.
29 Penney 1968, 297.
30 Hill 2013, 204.
31 *Nuclear Engineering*, August 1956, wall chart DIMPLE; Collins et al. 1963, 1.
32 *Electrical Review*, 30 November 1956, 1013–14: Oliphant and Penney 1968, 170; Hance 2006, 81.
33 *Electrical Review*, 15 March 1957, 488.
34 AEE 1977, 3.
35 Dounreay Site Restoration Ltd 2010, 17.
36 *Electrical Review*, 30 January 1956, 138; Echo 1996, 7.
37 *Electrical Review*, 30 January 1956, 138.
38 Oliphant and Penney 1968, 172; Paterson 2008, 45; Hance 2006, 145–7.
39 *Electrical Review*, 31 January 1958, 221–4.
40 Hawkings 2000, 24.
41 *Electrical Times*, 30 January 1958, 164.
42 Gibbons and Jay 1960, 32.
43 *BEAMA*, May 1958, 83.
44 *Electrical Review*, 31 January 1958, 223; Hance 2006, 149.
45 *The Electrician* 30 July 1948, 317; Hance 2006, 46.
46 *Nuclear Energy Engineer*, 1959, 13, issue 134, 369–70, 377; *Electrical Times*, 30 January 1958, 165.
47 *Electrical Review*, 20 November 1959, 703.
48 *The Electrician*, January 1948, 259.
49 *Nature*, 23 February 1957, 400.
50 www.chilton-computing.org.uk/ral/literature/drawings.htm (acc. 19 October 2023).
51 Lavington n.d., 20.
52 Jay 1955, 43–53.
53 Echo 1996, 10; Hance 2006, 188–96.
54 Hansard, HC Deb., vol. 733, cols 1202–11 (8 August 1966).
55 TNA: WO 32/20339; The Redfern Inquiry 2010, 222, 234, 236.
56 Paterson 2008, 109.
57 Dounreay Site Restoration Ltd 2010, 17–18.
58 www.royalnavyresearcharchive.org.uk/FAA-Bases/Dounreay.htm (acc. 14 August 2023).
59 Paterson 2008, 25–6.
60 Dounreay Site Restoration Ltd 2010, 19.
61 Laird 1957, 845.
62 Paterson 2008, 117.
63 *Electrical Review*, 17 May 1957, 905; *Electrical Review*, 26 October 1962, 676; Dounreay Site Restoration Ltd 2010, 19.
64 *Electrical Review*, 26 October 1962, 679.
65 Barrett 1958, 89, 92–3.
66 Paterson 2008, 32.
67 *Electrical Review*, 17 May 1957, 899.
68 *BEAMA*, August 1957, 95; Dounreay Site Restoration Ltd 2010, 21.
69 *BEAMA*, August 1957, 93–6.
70 Dounreay Site Restoration Ltd 2010, 19.
71 Ibid., 20; *Electrical Review*, 17 May 1957, 902–4, *BEAMA*, August 1957, 96.
72 Paterson 2008, 70.
73 *Electrical Review*, 17 May 1957, 904–5; *BEAMA*, August 1957, 97–8.
74 Paterson 2008, 45.
75 Walford and Thomas 1961, 553–88.
76 *Electrical Review*, 5 June 1964, 850.
77 Paterson 2008, 77.
78 *Electrical Times*, 16 May 1957, 763.
79 *Electrical Review*, 4 March 1966, 331; *Engineering and Boiler House Review*, March 1966, 90.
80 Paterson 2008, 69–70.
81 Jensen and Ølgaard 1995, 32.
82 Ibid., 5.
83 Dounreay Site Restoration Ltd 2010, 21.
84 *Electrical Review*, 27 August 1971, 298.
85 Jensen and Ølgaard 1995, 9–10.

86 Ibid., 13.
87 *Electrical Times*, 17 February 1966, 245; *Electrical Review*, 18 February 1966, 267, 241.
88 *Electrical Review*, 21 September 1979, 16–17.
89 Paterson 2008, 78; *Decommissioning Dounreay* 2009, 6.
90 Dounreay Site Restoration Ltd 2010, 22.
91 Hansard, HC Deb., vol. 571, cols 621–84 (30 May 1957); Fry 2015, 9.
92 *Electrical Review*, 18 January 1957, 102.
93 Dobinson 2000, 248.
94 Abbott 1958, 537–8; *Industrial Architecture*, September–October 1960, 351–8.
95 *Industrial Architecture*, September–October 1960, 356.
96 Fry 2015, 76–8.
97 *Electrical Times*, 22 September 1960, 430.
98 *Industrial Architecture*, September–October 1960, 354–5.
99 Fry 2015, 82.
100 Ibid., 79.
101 Miller et al. 2003; Fry 2015, 78.
102 Fry 2015, 13.
103 Ibid., 15–16.
104 *Nuclear Energy Engineer*, September 1959, 440; *BEAMA*, August 1959, 87.
105 Fry 2015, 53.
106 *The Engineer*, 15 February 1963, 315; Fry 2015, 59.
107 *Electrical Review*, 24 March 1961, 531; Fry 2015, 54.
108 Fry 2015, 54.
109 *Electrical Review*, 16 September 1960, 472; *Electrical Review*, 23 September 1960, 542; Fry 2015, 52.
110 Penney 1968, 300.
111 *Industrial Architecture*, 1959, 2, issue 5, 350; *Industrial Architecture*, September/October 1960, 358.
112 *Electrical Times*, 3 July 1969, 52.
113 Hansard, HC Deb., vol. 733, cols 1202–11 (8 August 1966).
114 Fry 2015, 50.
115 *Electrical Review*, 4 January 1963, 26; Fry 2015, 56–7, 79.
116 *Electrical Review*, 3 April 1964, 519.
117 Fry 2015, 43.
118 UKAEA 1980.
119 *Electrical Review*, 31 March 1968, 319; Hill 2013, 299.
120 Herman 1976, 16–17; Moss 1977, 16–17.
121 Hill 2013, 302.
122 SODC 2014.
123 Close 1990, 37–45; https://ccfe.ukaea.uk/fusion-energy/the-tokamak/ (acc. 10 January 2024).
124 Woodward 1990, 855–73; www.gov.uk/government/news/jets-final-tritium-experiments-yield-new-fusion-energy-record (acc. 14 February 2024).
125 UKAEA 6 October 2022 'West Burton selected as home of STEP fusion plant', www.gov.uk/government/news/west-burton-selected-as-home-of-step-fusion-plant (acc. 12 July 2023).
126 *Engineering and Boiler House Review*, July 1965, 245.

Box 4.1

1 Jay 1955, 26–35.
2 Kitson 2007, 738.
3 Jefferson 1963, 243–64.
4 RIBA collection 63351.
5 Kitson 2007, 739.
6 Ibid., 740.
7 Ibid.
8 Ibid.
9 Ibid., 741–2.
10 www.rsc.org/news-events/articles/2005/02-february/grove-centre/ (acc. 21 July 2023).
11 IEE 1993, 56.

Box 4.2

1 *Electrical Review*, 31 January 1958, 224; *Nuclear Energy Engineer*, 1959, 12, issue 126, 397.
2 *Electrical Review*, 4 September 1959, 196–7; *Nuclear Energy Engineer*, 1959, 13, issue 138, 536.
3 *Electrical Review*; 24 February 1956, 293–4.
4 *Nuclear Energy Engineer*, 1959, 13, issue 138, 536.
5 *Electrical Review*, 9 October 1959, 441.
6 PLA Monthly 1963, 82; Edwards 1969, 12–20.
7 HSE 2005 'HSE Revokes the Nuclear Site Licence for ICI's Triga Reactor and Ends the Licensees' Period of Responsibility', www.hse.gov.uk/nuclear/periodic-safety-review/triga.htm (acc. 7 September 2011).
8 *Electrical Times*, 16 July 1964, 102.
9 HSE 1984 'Final Report on the De-commissioning and Delicensing of the Queen Mary College Research Reactor', www.hse.gov.uk/foi/releases/queenmary230906.htm (acc. 7 September 2011).
10 *New Scientist*, 1963, 37; *Electrical Times*, 22 April 1965, 619; *New Scientist*, 1965, 379.

Box 4.3

1 Grimes et al. 2014, 69.
2 Jones 2019, 50, 54–5.
3 Ibid., 68–80.
4 Hill 2013, 327–8.
5 Ibid., 213–14, 326–7.
6 Penney 1968, 299; Williams 1980, 45; Hill 2013, 325–6.
7 EACRP 1964, 45.
8 Hill 2013, 329–31; Grimes et al. 2014, 25, 78; Jones 2019, 190–3.
9 www.baesystems.com/en-uk/heritage/dreadnought-submarine (acc. 3 January 2024).
10 Hill 2013, 321.
11 Grimes et al. 2014, 74.
12 Ibid., 24, 51.
13 www.radiationworks.com/nuclearships.htm (acc. 15 November 2023).
14 Justo and dos Santos 2000, 21–49.
15 Drollinger et al. 2000, 15.
16 www.nasa.gov/tdm/space-nuclear-propulsion/ (acc. 16 April 2024).
17 Drollinger et al. 2000, 16–17.
18 Hacker 1995, 85–98; Drollinger et al. 2000, 7.

Chapter 5

1 Hansard, HC Deb., vol. 627, cols 953–55 (27 June 1960).
2 Arnold 2007, 30.
3 Jay 1956, 13–14.
4 Ibid., 15–18.
5 Hall 1963, 5.
6 Oliphant and Penney 1968, 168.
7 Arnold 2007, 21.
8 *Electrical Review*, 19 October 1956, 693–5; *Engineering and Boiler House Review*, January 1956, 8; Jay 1956, 49–51; Stretch 1957, 419.
9 Arnold 2007, 28.
10 *Engineering and Boiler House Review*, May 1959, 143.
11 *Electrical Review*, 8 May 1959, 846–7.
12 *Electrical Review*, 22 June 1962, 1063, 1066.
13 Hall 1963, 33.
14 CEGB Barnwood 1986, 69.
15 Hall 1963, 45.
16 *Electrical Review*, 4 March 1960, 421–3.
17 *Electrical Times*, 3 November 1960, 677–8; *Electrical Review*, 4 November 1960, 779–80; *Electrical Review*, 6 July 1962, 3–10.

18 *Engineering and Boiler House Review*, March 1957, 86.
19 *The Engineer*, 8 February 1957, 217.
20 *Engineering and Boiler House Review*, March 1957, 86; *The Engineer*, 8 February 1957, 217; *Electrical Times*, 24 September 1964, 438.
21 *Nuclear Engineering*, October 1957, 423; *Electrical Times*, 12 July 1962, 46–7.
22 *Electrical Review*, 20 September 1957, 503.
23 *Nuclear Energy Engineer*, October 1959, 489–90, 514.
24 CEGB 1972, 18.
25 *Electrical Times*, 30 September 1965, 501.
26 *Electrical Review*, 2 November 1962, 691–6.
27 Taylor 1989, 54.
28 *Engineering and Boiler House Review*, February 1963, 56.
29 *Electrical Times*, 10 October 1968, 513.
30 *Electrical Times*, 1 May 1972, 66–7.
31 *Electrical Review*, 20 November 1981, 27.
32 Hall 1963, 35–7.
33 Ibid., 78–9.
34 *Electrical Review*, 17 June 1966, 905–6.
35 *Electrical Times*, 1 May 1972, 66–7.
36 *Electrical Review*, 25 November 1955, 1018; *Engineering and Boiler House Review*, May 1959, 144–5.
37 Hall 1963, 19.
38 *Electrical Times*, 27 March 1958, 505–7; *Electrical Review*, 28 March 1958, 589–90.
39 *Engineering and Boiler House Review*, September 1962, 308–9.
40 *Electrical Times*, 15 July 1963, 93.
41 *Electrical Review*, 7 July 1961, 10–12.
42 *Electrical Review*, 26 May 1961, 926.
43 Wooldridge and Druce 2019, 200.
44 BNFL n.d. a and b.
45 World Nuclear News 2022.
46 Hall 1963, 1.
47 Robin 1955, 1108.
48 *Nuclear Power*, September 1962, 68–9.
49 Oswald 1957, 3–7.
50 Viney and Brocklesby 1956, 528–33.
51 *Electrical Review*, 28 March 1958, 589–90.
52 *Concrete and Construction Engineering*, December 1960, 455–6.
53 *Concrete and Construction Engineering*, February 1960, 89.
54 www.powerstations.uk/littlebrook-fred/ (acc. 13 October 2023).
55 TNA: POWE 14/618 'Design and Construction of Power Station Economy of Building Materials', 30 Aug 1952 para 4 draft letter.
56 *Official Architecture and Planning*, December 1953, 589–90; TNA: POWE 14/617, piece 250, 1.
57 TNA: POWE 14/617.
58 *Official Architecture and Planning*, December 1953, 589–90.
59 Public and General Acts and Measures 1957, 5 and 6 Eliz, c. 48, *Electricity Act*, para 37.
60 *Nuclear Engineering*, April 1956, 1.
61 *Nuclear Engineering*, June 1958, 132.
62 Nuclear Electric 1992, 5.
63 *The Builder*, 12 October 1956, 616.
64 Gill 1964, 126.
65 *Electrical Times*, 14 February 1957, 261.
66 SSEB 1960, 4.
67 *Architect and Building News*, 13 June 1957, 774–81.
68 Crowe 1960, 3–6.
69 *Official Architecture & Planning*, November 1957, 557.
70 *Architect and Building News*, 13 June 1957, 774–81.
71 *Electrical Review*, 22 June 1962, 1063.
72 TNA: POWE 14/1244 Trawsfynydd: Public Inquiry, 55.
73 *Engineering and Boiler House Review*, October 1962, 334–9; *Oil Engine and Gas Turbine*, November 1962, 235; Hall 1963, 7.
74 *Official Architecture and Planning*, November 1957, 557.
75 SSEB 1960, 40.
76 *Electrical Review*, 20 July 1956, 122; *Nuclear Energy Engineer*, September 1958, 282.
77 Jellicoe 1961, 527–31.
78 *Institute of Landscape Architects*, November 1968, 14–17.
79 *Electrical Review*, 21 February 1958, 373.
80 TNA: POWE 14/1189 Notes prepared by Miss Sylvia Crowe.
81 *Official Architecture and Planning*, May 1969, 547; Purdy 1999, 122–5; Campbell 1999, 69–73.
82 *Engineering and Boiler House Review*, June 1965, 206–9.
83 Purdy 1999, 125–6; AR CRO PF/A/14 CEGB.
84 Cherry and Penny 1986, 207–9.
85 Hecht 1998, 60–1, 95.
86 Cook 1961, 778.
87 Taylor 2016, 15.

Box 5.1

1 TNA: AB 1/68 drwg10545/5.
2 Henderson 1956, 509.
3 *Electrical Times*, 10 July 1969, 5.
4 *Electrical Review*, 18 October 1957, 697–9; *Electrical Review*, 4 March 1960, 421.
5 *Electrical Review*, 28 August 1959, 118–19.
6 *Nuclear Engineering*, 1964, 8, issue 87, 288; Sten Architecture 2015, 26.
7 www.centronic.co.uk/history (acc. 3 October 2023).
8 *Electrical Review*, 22 June 1962, 1063; Hall 1963, 30.
9 *Electrical Review*, 19 October 1956, 693.
10 *Electrical Times*, 3 November 1960, 677–8.
11 *Electrical Times*, 10 July 1969, 5.
12 *Electrical Review*, 1960, 484–6.

Box 5.2

1 McAlone 2020, 46.
2 www.therailwayhub.co.uk/67072/from-the-archive-operation-smash-hit-84/ (acc. 2 January 2024).
3 McAlone 2020, 32–3, 50.

Box 5.3

1 https://inl.gov/experimental-breeder-reactor-i/ (acc. 29 September 2023); Josephson 1999, 27.
2 Reflector 1958, 107.

Chapter 6

1 *Electrical Review*, 10/17 April 1981, 9.
2 *Engineering and Boiler House Review*, April 1963, 133.
3 *Electrical Review*, 17 November 1960, 781.
4 Ibid.
5 *Electrical Times*, 9 June 1960, 928; *Engineering and Boiler House Review*, April 1963, 135.
6 *Electrical Review*, 10 June 1960, 1109–10.
7 *Electrical Review*, 10/17 April 1981, 9.
8 *Electrical Times*, 15 February 1962, 259; *Nuclear Energy Engineer*, April 1959, 189.
9 Moore 1964, 3–25.

10 Ministry of Power, 1964.
11 *Electrical Times*, 5 August 1965, 196.
12 Wooldridge and Druce 2019, 58–60.
13 CEGB Barnwood 1986, 72.
14 *Nuclear Engineering*, July 1967, 538.
15 *Electrical Times*, 18 November 1968, 729.
16 *Electrical Times*, 6 November 1970, 5.
17 *Electrical Times*, 9 November 1972, 19; *Electrical Review*, 2 June 1978, 5; Moss 1981, 5.
18 *Electrical Review*, 7 January 1983, 9.
19 *Electrical Times*, 18 May 1967, 838.
20 Young 1967, 536–7; *Electrical Times*, 2 January 1969, 8.
21 Taylor et al. 1971, 506; Taylor et al. 1972, 214.
22 Taylor et al. 1971, 508–9.
23 Benett 1967, 534–5.
24 Taylor et al. 1972, 217.
25 Taylor et al. 1971, 502, 504.
26 Duffy 1967, 524.
27 CEGB Barnwood 1986, 73, 467.
28 *Electrical Times*, 10 July 1969, 5.
29 Cave 1967, 527.
30 Waine 1967, 531–4.
31 *Electrical Review*, 26 May 1972, 731; Batten 1973, 275–6.
32 *Electrical Review*, 1 July 1966, 5; *Engineering and Boiler House Review*, August 1966, 247.
33 *Electrical Times*, 9 February 1967, 245.
34 *The Engineer*, 20 April 1962, 700.
35 *Electrical Review*, 10 March 1967, 355; *Electrical Review*, 13 February 1976, 5.
36 *Electrical Review*, 10 September 1976, 9.
37 *Electrical Times*, 9 March 1967, 383; *Electrical Review*, 26 June 1970, 933.
38 *Electrical Times*, 26 September 1968, 445.
39 *The Engineer*, 25 October 1968, 629.
40 *Electrical Review*, 28 November 1969, 785.
41 *The Engineer*, 27 September 1968, 462.
42 *The Engineer*, 10 February 1967, 220.
43 *Electrical Review*, 10 September 1976, 9.
44 *Electrical Review*, 26 June 1970, 933.
45 *The Engineer*, 10 February 1967, 221; *The Engineer*, 27 September 1968, 462.
46 *Engineering and Boiler House Review*, May 1967, 140–1.
47 *Electrical Times*, 26 December 1968, 871.
48 *Electrical Review*, 10 September 1976, 9.
49 *Electrical Times*, 28 July 1966, 151; *Electrical Times*, 6 July 1967, 28; *Electrical Review*, 13 February 1976, 5.
50 *Electrical Times*, 19 October 1967, 626.
51 *Electrical Times*, 6 July 1967, 28.
52 Cooper 1982, 5.
53 *Electrical Times*, 16 March 1967, 444.
54 *Electrical Times*, 13 October 1967, 529.
55 *Engineering and Boiler House Review*, August 1966, 247.
56 *Electrical Times*, 24 October 1968, 572; *Concrete*, April 1969, 146.
57 *Electrical Review*, 30 August 1968, 300.
58 Moss 1981, 5.
59 *Electrical Review*, 6 December 1968, 837; *Electrical Review*, 21 July 1978, 9.
60 *Electrical Review*, 3 July 1970, 4.
61 *Electrical Times*, 6 August 1970, 6.
62 *The Engineer*, 29 November 1968, 823–4.
63 *Electrical Review*, 6 September 1968, 335; *Electrical Times*, 24 October 1968, 572; *The Engineer*, 29 November 1968, 823–4.
64 *The Engineer*, 29 November 1968, 823–4.
65 *Electrical Times*, 16 June 1966, 905; *Electrical Times*, 24 October 1967, 173; *Electrical Review*, 3 February 1967, 157.
66 *Electrical Times*, 6 November 1969, 5.
67 *Electrical Review*, 11 December 1970, 826.
68 *Electrical Review*, 3 December 1971, 831.
69 *Electrical Review*, 9 January 1981, 5.
70 *Electrical Times*, 16 October 1969, 7.
71 Hecht 1998, 297–9.
72 Grimston and Nuttall 2013, 17.
73 Baker 2023, para 31.
74 Ibid.
75 Wooldridge and Druce 2019, 80.
76 Ibid., 86.
77 Ibid., 88, 96.
78 CEGB Barnwood 1986, 77, 433.
79 *Electrical Review International*, 6 July 1979, 7; *Electrical Review*, 23 October 1981, 5.
80 *Electrical Review*, 1 August 1980, 7.
81 *Electrical Review*, 9 January 1981, 5.
82 CEGB Barnwood 1986, 436–7.
83 *Electrical Review*, 9 January 1981, 5; HSE 2001, 4.
84 SSEB n.d., 3, Hinton and Holford 1960, 182–88, 196.
85 Ibid.
86 Ibid., 17–19; *Electrical Review*, 9 January 1981, 5.
87 SSEB n.d., 16.
88 Ibid., 10.
89 Ibid.
90 Wooldridge and Druce 2019, 202.
91 Ibid., 203.
92 Ibid., 207.
93 Jordan and Mann 1990, 13.
94 Wooldridge and Druce 2019, 208–9.
95 https://shepleyengineers.co.uk/case-studies/thorp-receipt-and-storage-building-cladding/ (acc. 8 November 2023).
96 Jordan and Mann 1990, 7–13.
97 Field visits by author, December 1998 and July 2010.
98 Wooldridge and Druce 2019, 208–10.
99 Lobb 1972, 214.
100 Ibid.
101 CEGB Barnwood 1986, 530–2.
102 *Electrical Times*, 16 May 1968, 835; *Electrical Times*, 6 November 1969, 5.
103 *Electrical Times*, 5 March 1971, 27; CEGB Barnwood 1986, 542.
104 *Electrical Review*, 24 August 1973, 246.
105 CEGB Barnwood 1986, 539.
106 *Electrical Review*, 14 May 1982, 5.
107 Brookes 1982, 14–15; Carter 1987; *Electrical Review*, October 1989, 13.
108 Carter 1981, 7.
109 *Electrical Review*, 16–23 May 1986, 6.
110 *Electrical Review*, 3 October 1986, 5.
111 CEGB Barnwood 1986, 557; Cogle 1987, 16–17; Baker 2023, para 37.
112 *Electrical Review*, 21/28 August 1981, 9.
113 Cogle 1987, 16–17; Baker 2023, para 37.
114 CEGB Barnwood 1986, 557.
115 www.building.co.uk/focus/sizewell-stories/3069773.article (acc. 3 October 2023).
116 Youngman 1985, 34.
117 *Electrical Review*, 13/20 April 1984, 4.
118 *Electrical Review*, 27 February 1987, 7; CEGB Barnwood 1986, 644.
119 *Electrical Review*, 29 January 1982, 17.
120 CEGB Barnwood 1986, 570.
121 Ibid., 584–6.

Box 6.1

1 Hecht 1998, 201–4.
2 *Heating* 1958, 246–9; TNA: BP 2/72.
3 Hall 1963, 11.
4 Letter from The Institution of Mechanical Engineers, 4 July 1957, in author's possession.
5 Ross 2023, 208.
6 'Heysham Power stations visitor centre', www.hparch.co.uk/leisure/projects/heysh.html (acc. 7 March 2005).

Chapter 7

1 www.gov.uk/government/organisations/nuclear-decommissioning-authority/about (acc. 8 February 2024).
2 www.highlifehighland.com/nucleus-nuclear-caithness-archives/ (acc. 8 February 2024).
3 Dounreay Site Restoration Limited 2010; Croft 2019, 103–9.
4 www.nps.gov/mapr/index.htm (acc. 12 February 2024).
5 www.kirtland.af.mil/News/Article-Display/Article/390159/national-atomic-museum-has-new-name-new-home/ (acc. 11 February 2024).
6 www.energy.gov/ne/articles/9-notable-facts-about-worlds-first-nuclear-power-plant-ebr-i (acc. 12 February 2024).
7 Rindzevičiūtė 2021, 839–62.
8 Philips 2013, 20.
9 Mills 2023, 3–4; UK Parliament, 'AWE Aldermaston and AWE Burghfield', Question for Ministry of Defence, UIN 45, tabled on 10 May 2022. https://questions-statements.parliament.uk/written-questions/detail/2022-05-10/45 (acc. 8 February 2024).
10 https://assets.publishing.service.gov.uk/media/65c0e7cac43191000d1a457d/6.8610_DESNZ_Civil_Nuclear_Roadmap_report_Final_Web.pdf (acc. 8 February 2024).

Select bibliography and further reading

Primary sources

The National Archives

Records of the Ministry of Supply/United Kingdom Atomic Energy Research Authority

AB 1/68 M.S. Factory, Valley, Flintshire Metro-Vick Models Ltd 1941–43.
AB 6/326 Sellafield (Windscale) General April 1947–December 1948.
AB 8/134 Housing and hostels – Housing Windscale general 1947–1949.
AB 8/272 Housing Windscale Whitehaven Corporation.
AB 8/557 History of the Windscale Production Piles 1945–51.
AB 16/1890 Woolwich Common site: general 1955–1957.
AB 16/916 Foulness: Explosion damage; claims 1953–4.
AB 16/2228 Search for non-nuclear weapons test site and acquisition of Orford Ness.
AB 16/2975 Orford Ness Phase III experimental test facilities 1959–1966.
AB 44/6 Woolwich Common site: correspondence 1963–1964.
AB 46/10 Manpower problems of individual establishments: closure of Woolwich Common, London 1-1-1963 31-21-1963.
AB 58/8 Windscale Works a preliminary brief survey carried out by the Department of Atomic Energy, prior to taking over the late ROF Sellafield for their own use August 1947.

Records of the Air Ministry

AIR 2/13689 Fuzing systems for atomic bombs.

Records of the Ministry of Aviation

AVIA 65/1258 Atomic Weapons Research Establishment: minutes on vibration test rig 1955–57.

Records of the Royal Fine Art Commission

BP 2/72 Demolition of 11 and 12 Charles II Street and the erection of a new building: headquarters of the Atomic Energy Authority. 1954–56.
BP 2/145 Windscale atomic station.

Records of the Atomic Weapons Research Establishment

ES 1/246 Atomic Weapons Research Establishment: Rowley Collection move to Aldermaston from Fort Halstead: travel allowances 1950.
ES 1/330 Development of Foulness 1945–51.
ES 1/331 Development of Foulness 1952–1956.
ES 1/267 Buildings for HER project: Shoeburyness.
ES 1/269 Buildings for HER Project: Shoeburyness and Foulness.
ES 1/272 Works Services: Woolwich Arsenal 1947–54.
ES 1/691 1956–57 Woolwich Common extensions: buildings and extensions 1956–1957.
ES 1/1420 1953–55 Woolwich Common use by AWRE 1953–1955.

Records of the Ministry of Power

POWE 14/617 Siting and design, economy in construction of generating stations including appointment of committee under Sir H. Beaver.
POWE 14/618 Power station individual siting and design.
POWE 14/1189 Nuclear power stations: Siting at Trawsfynydd, Merionethshire.
POWE 14/1244 Trawsfynydd: Public Inquiry minutes.

Records of the War Office

WO 32/20339 Woolwich Review Committee: replies to questionnaires and minutes of other meetings to discuss future of Woolwich estate 1963–65.

Defence Science and Technology Laboratory Fort Halstead (on-site microfiche)

F16 AB 1/1 5839/21A Bomb Chamber 20 July 1947.
F16 Fort Halstead H.E. Research Bomb Chamber, Lightning Protection Installation 12.9.47.
F17 AB 2/1 5839/32A Fort Halstead H.E. Research Detonation Laboratory August 1947.

National Trust, Bury St Edmunds

Penney, W. G. letter 23 March 1955 source AWE file 0239 1a.
UKAEA 1962 Lab2 Orford Ness general arrangement of centrifuge installation in Lab 2 Architect G W Dixon 24 6 62.
UKAEA 1963 SW/S7851/Lab2/AD18339 Lab No. 2 Centrifuge Details of plant room extension 24 9 63.

University of Reading, Museum of Rural Life

AR CRO PF/A/14 CEGB Wylfa Nuclear Power Station Landscape Report No. 2 1962.

London Metropolitan Archives

GLC/AR/BR/26/WO/M/241-2: LBGPA.

Secondary sources

Abbott, L. 1958 'Winfrith Atomic Establishment anticipated impact on Dorset', *Official Architecture and Planning*, 21, issue 11, 517–18.

AEE 1977 *Nuclear Power Reactors* (AEE Winfrith booklet).

Andrade, E. N. da C. 1964 *Rutherford and the Nature of the Atom* (New York: Anchor).

Arnold, L. 2001 *Britain and the H-Bomb* (Basingstoke: Palgrave).

Arnold, L. 2007 *Windscale 1957: Anatomy of a Nuclear Accident*, 3rd edn (Basingstoke: Palgrave).

AWRE n.d. *Environmental Testing* (AWRE Orford Ness brochure *c.* 1968).

Aylen, J. 2015 'First waltz: Development and deployment of Blue Danube, Britain's post-war atomic bomb', *International Journal for the History of Engineering and Technology*, 85, issue 1, 31–59.

Baker, K. 2023 'Nuclear power in Britain: A series of successful failures', *International Review of Public Policy*, 5, issue 1, 26–43.

Barrett, N. T. 1958 'Housing the Dounreay reactor', *Structural Engineer*, 36, 85–97.

Batten, T. F. 1973 'Successful run up of Dungeness B 660MW turbine generator under computer control', *Electrical Review*, 23 February, 275–6.

Benett, J. C. 1967 'Dungeness B reactor core design', *Nuclear Engineering*, 12, 534–5.

Bone, P., Nichol, K., Pearson, N., and Peters, T. S. 2006 *The Valley Site, Rhydymwyn, Flintshire: Historic Environment Management Plan*, vol. 1, Birmingham Archaeology Report No. 1377.

Bowyer, M. J. F. 1983 *Action Stations 6. Military Airfields of the Cotswolds and the Central Midlands* (Cambridge: Patrick Stephens).

BNFL n.d. a *Redundant Magnox Storage and Decanning Facility B30* (BNFL leaflet).

BNFL n.d. b *Intermediate Level Waste Storage Facility B38* (BNFL leaflet).

Brookes, L. 1982 'The case for Sizewell "B"', *Electrical Review*, 211, issue 13, 14–15.

Bruce, K. 2002 *Bradwell Power Station: 40 Years of Power* (Bradwell: Bradwell Power Station).

Buckley, R. 2004 *Nuclear Energy Technology, Risk and Climate Change*, Understanding Global Issues 129 (Cheltenham: Understanding Global Issues).

Campbell, D. 1999 'Forestry', in G. Collens and W. Powell, *Sylvia Crowe*, LDT Monograph No. 2, 69–107.

Carroll, S. J. 2010 '"Fill the Jails": Identity, Structure and Method in the Committee of 100, 1960–1968' (DPhil thesis, University of Sussex), http://sro.sussex.ac.uk/id/eprint/6910/1/Carroll,_Samantha_Jane.pdf (acc. 9 January 2023).

Carter, J. 1981 'Sizing up the risk of nuclear power', *Building Design*, July 24, 7.

Carter, J. 1987 'Legal opinion', *Building Design*, February, 822.

Cathcart, B. 1994 *Test of Greatness: Britain's Struggle for the Atom Bomb* (London: John Murray).

Cathcart, B. 2004 *The Fly in the Cathedral* (London: Penguin).

Cave, L. 1967 'Dungeness B safety aspects', *Nuclear Engineering*, 12, 526–8.

CEGB 1972 *Dungeness 'A' Nuclear Power Station* (London: CEGB).

CEGB Barnwood 1986 *Advances in Power Station Construction* (Oxford: Pergamon).

Cherry, G. E., and Penny, L. 1986 *Holford: A Story in Architecture, Planning and Civic Design* (London: Mansell).

Clark, R. W. 1961 *The Birth of the Bomb* (London: Scientific Book Club).

Clark, R. W. 1980 *The Greatest Power on Earth: The Story of Nuclear Fission* (London: Sidgwick and Jackson).

Close, F. 1990 *Too Hot to Handle: The Story of the Race for Cold Fusion* (London: W. H. Allen).

Cocroft, W. D. 2006 *England's Atomic Age Strategy on the Historic Industrial Environment Report Desk Top Investigation and Assessment* Part 1, English Heritage RDRS 3-2006.

Cocroft, W. D. 2010 *Fort Halstead, Dunton Green, Sevenoaks, Kent: A Brief Assessment of the Role of Fort Halstead in Britain's Early Rocket Programmes and on the Atomic Bomb Project*, English Heritage RDRS 49-2010.

Cocroft, W. D., and Alexander, M. 2009 *Atomic Weapons Research Establishment, Orford Ness, Suffolk: Cold War Research and Development Site*, English Heritage RDRS 10-2009.

Cocroft, W. D., and Newsome, S. 2009 *Atomic Weapons Research Establishment, Foulness, Essex: Cold War Research and Development Site*, English Heritage RDRS 13-2009.

Cocroft, W. D., and Thomas, R. J. C. 2003 *Cold War Building for Nuclear Confrontation 1946–89* (Swindon: English Heritage).

Cogle, J. 1987 'Sizewell – A decision without a policy', *Electrical Review*, 220, issue 4, 16–17.

Collens, G., and Powell, W. 1999 *Sylvia Crowe*, LDT Monograph No. 2 (Reigate: LDT).

Collins, A. G., Taylor, W. E., and Wells, G. M. 1963 *The First Core in DIMPLE at A.E.E. Winfrith – Experimental Results* (AEE Winfrith typescript report W.5877).

Cook, W. 1961 'Reactor development in the United Kingdom', *Electrical Review*, 17 November, 771–8.

Coolidge, M. 1996 *The Nevada Test Site: A Guide to America's Proving Ground* (Culver City: Center for Land Use Interpretation).

Cooper, G. 1982 'Electricity bills close Invergordon smelter', *Electrical Review*, 210, issue 1 (8 January), 5.

Croft, A. 2019 'Dounreay heritage strategy: White heat of heritage', in ICOMOS, *Kernkraftwerke. Denkmalwerte und Erhaltungsschancen*, 103–9.

Crowe, S. 1960 'Power and landscape', *Journal of the Landscape Institute*, 52, 3–6.

Daily Mail 1954 'Atomics in the Service of Mankind'.

Davey, H. G. 1994 'Reminiscences of an atom pioneer', 1/1/94 (internal BNFL document).

Davies, H. (ed.) 2012 *Sellafield Stories: Life with Britain's First Nuclear Power Plant* (London: Constable).

Decommissioning Dounreay 2009 'The £1.99 cleaning product that gets rid of plutonium stains', *Decommissioning Dounreay*, 7 (Autumn), 6.

Department for Energy Security and Net Zero 2024 *Civil Nuclear Roadmap to 2050*, https://assets.publishing.service.gov.uk/media/65c0e7cac43191000d1a457d/6.8610_DESNZ_Civil_Nuclear_Roadmap_report_Final_Web.pdf (acc. 8 February 2024).

Dick, D. R. R. 1954 'The design and construction of the nuclear reactor buildings at Windscale, Sellafield', *Structural Engineer*, 32, issue 11, 287–303.

Dick, D. R. R. 1955 'The civil engineer and Britain's atomic factories', *Proceedings of the Institution of Civil Engineers*, 4, issue 4, 514–36.

Dobinson, C. 2000 *Fields of Deception: Britain's Bombing Decoys of World War II* (London: Methuen).

Dounreay Site Restoration Limited 2010 *Dounreay Heritage Strategy: Delivering a Cultural Legacy through Decommissioning* (typescript).

Drollinger, H., Goldenberg, N., and Beck, C. M. 2000 *An Historical Evaluation of the R-MAD Building in Area 25 for Planned Activities Associated with the Environmental Management, Decontamination and Decommissioning Program, Nevada Test Site, Nye County, Nevada* (Las Vegas: Desert Research Institute).

Drollinger, H., Jones, R., and Beck, C. M. 2000 *Historical American Engineering Record Nevada Test Site, PLUTO Facility, Disassembly Building 2201, HAER No. NV-32-A* (San Francisco: Historical American Engineering Record).

Duffy, E. P. 1967 'Dungeness B', *Nuclear Engineering*, 12, 524–5.

eCHO 1996 *Harwell 1946–1996: Special Anniversary Edition* (Abingdon: UKAEA).

Edwards, J. 1969 'Courses in nuclear science and technology at the Royal Naval College, Greenwich', *Institution of Mechanical Engineers*, 183 (3F), 12–20.

Elwall, R. 2000 *Building a Better Tomorrow* (London: Wiley-Academy).

EACRP 1964 *List of Research Reactors and Critical Facilities* (Paris: European American Committee on Reactor Physics).

Francis, P., and Crisp, G. 2008 *Military Command and Control Organisation*, vol. 1: *The Royal Navy*, Historic England Research Report 109-2008.

Fry, P. 2015 *Remembering AEE Winfrith: A Technological Moment in Time* (Stroud: Amberley).

Garratt, J. 2016 'Atomic Spaces in Northwest England 1945–1957' (Master of Science thesis, University of Manchester).

Gibbons, G., and Jay, K. 1960 *British Nuclear Reactors* (London: Chatto and Windus).

Gibson, J. H. 1996 *Nuclear Weapons of the United States: An Illustrated History* (Atglen, PA: Schiffer).

Gill, S. 1964 *Structures for Nuclear Power* (London: C. R. Books).

Gowing, M. 1964 *Britain and Atomic Energy 1939–1945* (London: Macmillan).

Gowing, M. 1974 *Independence and Deterrence: Britain and Atomic Energy, 1945–1952*, vol. 2: *Policy Execution* (London: Macmillan).

Gowing, M. 1987 *How Nuclear Power Began* (Southampton: University of Southampton).

Griffiths, N. 1984 *R.A.R.D.E. Fort Halstead: A Short History* (typescript).

Grimes, R., Ions, S., and Sherry, A. 2014 *Royal Navy Nuclear Reactor Test Facility Review*, https://assets.publishing.service.gov.uk/media/5a8238e840f0b62305b9321c/2015-03573.pdf (acc. 2 January 2024).

Grimston, M. C., and Nuttall, W. J. 2013 *The Siting of UK Nuclear Power Installations* (Cambridge: Energy Policy Research Group, University of Cambridge), www.jstor.org/stable/resrep32432 (acc. 25 September 2023).

Hacker, B. 1995 'Whoever heard of nuclear ramjets?', *ICON*, 1, 85–98.

Hall, J. 1963 *The Story of the Construction of Berkeley Nuclear Power Station* (London: Leonard Hill).

Hance, N. 2006 *Harwell: The Enigma Revealed* (Oxford: Enhance).

Hansard, HC Deb., vol. 415, cols 38–9 (29 October 1945) 'Atomic Energy (Government Research Establishment)'.

Hansard, HC Deb., vol. 477, col. 115 (19 June 1950) 'Mobile Labour Force (Capenhurst)'.

Hansard HC Deb., vol. 498, col. 212 (8 April 1952) 'Capenhurst Atomic Site (Hostel)'.

Hansard, HC Deb., vol. 508, col. 138 (2 December 1952) 'Fitzharris House, Abingdon (Demolition)'.

Hansard, HC Deb., vol. 571, cols 621–84 (30 May 1957) 'Winfrith Heath Bill'.

Hansard, HC Deb., vol. 627, cols 953–55 (27 June 1960) 'Nuclear Power Programme'.

Hansard, HC Deb., vol. 722, cols 624–8 (9 December 1965) 'Atomic Energy Authority Capenhurst Plant'.

Hansard, HC Deb., vol. 733, cols 1202–11 (8 August 1966) 'Atomic Energy Authority Establishment, Bracknell'.

Harwood, E. 2010 *England's Schools: History, Architecture and Adaptation* (Swindon: English Heritage).

Hawkings, D. J. 2000 *Keeping the Peace: The Aldermaston Story* (Barnsley: Leo Cooper).

Heating 1958 'High-velocity air conditioning for a London office building', *Heating*, 20, issue 153, 246–9.

Hecht, G. 1998 *The Radiance of France: Nuclear Power and National Identity after World War II* (Cambridge, MA: MIT Press).

Henderson, W. A. 1956 'Building for atomic power plant', *Architects' Journal*, 11 October, 507–34.

Hennessy, P. 2003 *The Secret State: Whitehall and the Cold War* (London: Penguin).

Hennessy, P. 2007 *Cabinets and the Bomb* (Oxford: Oxford University Press).

Herman, R. 1976 'Debate on heavy water reactors raises steam', *Electrical Review*, 199, issue 20 (19 November), 16–17.

Hill, C. N. 2013 *An Atomic Empire: A Technical History of the Rise and Fall of the British Atomic Energy Programme* (London: Imperial College).

Hinton, C., and Holford, W. 1960 'Power production and transmission in the countryside: Preserving amenities', *Journal of the Royal Society of Arts*, 108, 180–201.

Hughes, C. n.d. *Blue Danube: An Inside Story of Britain's First Atomic Bomb* (unpublished manuscript).

Hughes, J. 2002 *The Manhattan Project: Big Science and the Atom Bomb* (Duxford: Icon).

ICOMOS 2019 *Kernkraftwerke. Denkmalwerte und Erhaltungsschancen.*
ICOMOS Hefte des Deutschen Nationalkomitees LXVIII.
IEE 1993 *Nuclear power in the UK*, 2nd ed., 1993–4 (London: Public Affairs Board Institution of Electrical Engineers).
IAEA 1964 *Exponential and Critical Experiments*, Proceedings of a Symposium, Amsterdam, 2–6 September 1963, vol. III (Vienna: IAEA).
Jay, K. E. B. 1954 *Britain's Atomic Factories: The Story of Atomic Energy Production in Britain* (London: HMSO).
Jay, K. E. B. 1955 *Atomic Energy Research at Harwell* (London: Butterworths Scientific Publications).
Jay, K. E. B. 1956 *Calder Hall: The Story of Britain's First Atomic Power Station* (London: Methuen).
Jefferson, S. 1963 'Irradiation plant economics', in *Industrial Uses of Large Radiation Sources*, vol. 2 (Vienna: IAEA), 243–64.
Jensen, S. E., and Ølgaard, P. L. 1995 *Description of the Prototype Fast Reactor at Dounreay* (Roskilde: NKS).
Jellicoe, A. G. 1961 'The collaboration of civil engineers and landscape architecture from the point of view of the landscape architect', *Proceedings of the Institution of Civil Engineers*, 21, 527–35.
Jones, M. J. 2019 'The Development of Nuclear Propulsion in the Royal Navy, 1946–1975' (PhD thesis, University of Plymouth).
Jordan, G. W., and Mann, A. P. 1990 'THORP receipt and storage: Design and construction', *Structural Engineer*, 68, issue 1, 7–13.
Josephson, P. R. 1999 *Red Atom: Russia's Nuclear Power Program from Stalin to Today* (New York: W. H. Freeman).
Justo, L. F., and dos Santos, G. M. 2000 'The *Otto Hahn* nuclear ship and the German-Brazilian deals on nuclear energy – A case study in big science', *ICON*, 6, 21–49.
Kinsey, G. 1981 *Orford Ness – Secret Site: A History of the Establishment 1915–1980* (Lavenham: Terence Dalton).
Kitson, S. L. 2007 'Amersham radiochemistry to GE healthcare', *Journal of Labelled Compounds and Radiopharmaceuticals*, 50, 737–45.
Laird, M. 1957 'Architecture in the far north: A visit to the experimental atomic site at Dounreay', *Architects' Journal*, 125, 845–7.
Lavington, S. n.d. 'Ferranti Atlas 2 computers at AWRE Aldermaston and at the CAD Centre', https://curation.cs.manchester.ac.uk/atlas/docs/Atlas%202%20at%20AWRE%20and%20CAD%20Centre%20finalbb.pdf.
Litherland, S. 2006 *The Valley Site, Rhydymwyn, Flintshire: Historic Environment Management Plan*, vol. 2, Birmingham Archaeology Report No. 1377.
Lobb, H. 1972 in J. E. Taylor, R. F. Ready, and G. T. Williams, 'The superstructures of Dungeness B Nuclear Power Station – Discussion', *Structural Engineer*, 50, issue 5, 214.
Martin, A., and Harbison, S. A. 1986 *An Introduction to Radiation Protection*, 3rd edn (London: Chapman and Hall).
McAlone, D. 2020 *The Railways and Britain's Nuclear Industry* (Stamford: Key).
Miller, K. D., Parkinson, S. J., Cornell, R. M., and Staples, A. T. 2003 'Decommissioning of shielded facilities at Winfrith used for post irradiation examination of nuclear fuels and other active items', WM Symposium 23–27 February Tucson (Arizona), https://digital.library.unt.edu/ark:/67531/metadc782228/m1/1/.
Millington, C. F. 1971 *Design Study for an Environmental Test Building Erected at AWRE Orford Ness*, AWRE Report No. 0 34/71 (TNA ES4/1282).
Mills, C. 2023 *Replacing the UK's Nuclear Deterrent: The Warhead Programme*, House of Commons Library Research Briefing 9777, https://researchbriefings.files.parliament.uk/documents/CBP-9777/CBP-9777.pdf (acc. 8 February 2024).
Mills, E. A. 1951 *The Modern Factory* (London: Architectural Press).
Ministry of Power 1964 *The Second Nuclear Power Programme: Presented to Parliament by the Minister of Power by Command of Her Majesty, April 1964* (London: HMSO).
Ministry of Works 1949 *Harwell: The Radio-Chemical Laboratory* (London: MoW).
Moore, J. 1964 'Reactor physics development for advanced gas-cooled reactors', in *Exponential and Critical Experiments*, Proceedings of a Symposium, Amsterdam, 2–6 September 1963, vol. III (Vienna: IAEA), 32–4.
Moore, R. 2021 'Rudolf Peierls's "Outline of the development of the British tube alloy project": A 1945 account of the earliest UK work on atomic energy', *Nuclear Technology*, 207 (Sup1), S374–S379.
Moss, J. 1977 'Thumbs down for SGHWR, but is it to be AGR? Or PWR? Or both?', *Electrical Review*, 201, issue 6 (5 August), 16–17.
Moss, J. 1981 'It's all go, go, go at Hartlepool AGR', *Electrical Review*, 209, issue 3, 5.
NRPB 1989 *Living with Radiation* (London: HMSO).
Nature 1957 'Scientific and Industrial Research Co-Operation in the Commonwealth'. *Nature*, 179, 400.
Neufeld, M. J. 1995 *The Rocket and the Reich: Peenemünde and the Coming of the Ballistic Missile Era* (New York: Free Press).
New Scientist 1963 'Research reactor for Imperial College now ordered', *New Scientist*, 320, 37.
New Scientist 1965 'A college reactor goes critical', *New Scientist*, 442, 379.
Newsome, S., and Williams, A. 2011 *Woolwich Common, Woolwich, Greater London: An Archaeological Survey of Woolwich Common and its Environs*, Research Department Report 98-2009.
Nichol, K., and Litherland, S. 2006 'The Valley Site, Rhydymwyn, Flintshire Historic Environment Management Plan', vol. 3, *Birmingham Archaeology Report No. 1377.*
Norris, R. S., Burrows, A. S., and Fieldhouse, R. W. 1994 *British, French, and Chinese Nuclear Weapons*, Nuclear Weapons Databook V (Boulder, CO: Westview).
Nuclear Electric 1992 *Trawsfynydd Power Station* (Gloucester: Nuclear Electric).
Nuttall, W. J. 2005 *Nuclear Renaissance Technologies, and Policies for the Future of Nuclear Power* (Bristol: Institute of Physics).

Oliphant, M. L. E., and Penney, W. G. 1968 'John Douglas Cockcroft. 1897–1967', *Biographical Memoirs of Fellows of the Royal Society*, 14, 139–88, www.jstor.org/stable/769442 (acc. 7 September 2023).

Oswald, W. 1957 'Industry's contribution to nuclear power', *BEAMA Journal* (February), 3–7.

Paterson, W. 2008 *50 years of Dounreay* (Wick: North of Scotland Newspapers).

Penney, W. G. 1968 '1st Cockcroft Memorial Lecture: Sir John Cockcroft and Atomic Energy', *Journal of the British Nuclear Energy Society*, 7, issue 4, 295–302.

Philips, T. 2013 'Welcome to China's new theme park: The A-bomb testing site', *The Telegraph*, 8 November.

PLA 1963 'Nucleonics for the Navy', *Port of London Authority Monthly*, March, 82.

Purdy, I. 1999 'Landscape of power', in G. Collens and W. Powell, *Sylvia Crowe*, LDT Monograph No. 2, 121–32.

Pyne, K. 2006 'X marks the spot' (unpublished lecture, British Rocketry Oral History Project, Charterhouse).

Radford, J. E. 1961 *Nuclear Energy Simplified* (London: Macdonald).

Reader, W. J. 1975 *Imperial Chemical Industries: A History*, vol. 2: *The First Quarter Century, 1926–1952* (Oxford: Oxford University Press).

Reflector 1958 'Views on the news', *Electrical Review*, 17 January, 107.

Rindzevičiūtė, E. 2021 'Nuclear power as cultural heritage in Russia', *Slavic Review*, 80, issue 4 (Winter), 839–62.

Robin, K. J. 1955 'Industrial application of nuclear power', *Electrical Review* 24 June 1955, 1107–09.

Ross, L. M. 2019 '"Nuclear Fission and Social Fusion": The Impact of the Dounreay Experimental Research Establishment on Caithness, 1953–1966' (PhD thesis, University of the Highlands and Islands).

Ross, L. M. 2023 'Nuclear narratives: Rural modernity, identity and heritage in the Highlands and Islands', *Proceedings of the British Academy*, 256, 208–33.

Rowlands, J. 1990 'The development of the atomic bomb', *Proceedings of the Royal Air Force Historical Association*, 7, 13–18.

Smith, M. 1986 'SSEB tries to head off Torness protests', *Electrical Review*, 219, issue 1 (11/18 July), 11.

Smyth, H. D. 1945 *A General Account of the Development of Methods of Using Atomic Energy for Military Purposes under the Auspices of the United States Government 1940–1945* (London: HMSO).

SODC 2014 *Culham Science Centre Masterplan Consultation Draft Supplementary Planning Document* (South Oxfordshire District Council).

SSEB 1960 *Hunterston Nuclear Generating Station* (South of Scotland Electricity Board).

SSEB n.d. *Torness: The Way Forward* (South of Scotland Generating Board).

Sten Architecture 2015 *Design and Access Statement for a Full Planning Application for Development on Land at Newburgh Engineering Works, Bradwell*, https://portal.peakdistrict.gov.uk/system/download/30567898 (acc. 12 July 2023).

Stretch, K. L. 1957 'Nuclear power station operation', *Electrical Review* (8 March), 415–20.

Taylor, G. M. 1989 'Hunterston A and B: Operating the Magnox and AGR', *Nuclear News*, 32, issue 5, 52–62.

Taylor, J. 2001 'History of scientific computing at AWE – Part 1', *Discovery*, 2, 52–5.

Taylor, J. E., Ready, R. F., and Williams, G. T. 1971 'The superstructures of Dungeness B nuclear power station', *Structural Engineer*, 49, issue 11, 501–11.

Taylor, J. E., Ready, R. F., and Williams, G. T. 1972 'The superstructures of Dungeness B nuclear power station – Discussion', *Structural Engineer*, 50, issue 5, 213–17.

Taylor, S. 2016 *The Fall and Rise of Nuclear Power in Britain: A History* (Cambridge: UIT).

The Redfern Inquiry 2010 *The Redfern Inquiry into Human Tissue Analysis in UK Nuclear Facilities*, vol. 1: Report (London: The Stationery Office).

UKAEA 1980 *Winfrith Reactor: The 1000 MWe SGHWR Power Station* (AEE leaflet).

UKAEA n.d. *Project WAGR* (Windscale leaflet).

Vig, N. J. 1968 *Science and Technology in British Politics* (Oxford: Pergamon).

Viney, T. L. 1951 *A Report on the Construction of the Windscale Works, Sellafield, Cumberland, for the Division of Atomic Energy, Risley*, TNA: AB 7/1271.

Viney, T. L., and Brocklesby, R. S. 1956 'Notes on Calder Hall Reactor', *Architects' Journal* (11 October), 526–34.

Waine, P. 1967 'Dungeness B fuel handling', *Nuclear Engineering*, 12, 531–34.

Walford, J. G., and Thomas, A. F. 1961 'The equipment and methods used in British criticality laboratories', in *Criticality and Control in Chemical and Metallurgical Plant*, Karlsruhe Symposium 1961 (Paris: Organisation for Economic Cooperation and Development, European Nuclear Energy Agency) 553–88.

Watson-Munro, C. N. 1954 'Australian Atomic Energy Commission', *Nature*, 174, 1085.

Welsh, I. 2001 'Anti-nuclear movements: Failed projects or heralds of a direct action milieu', *Sociological Research Online*, 6, issue 3, https://doi.org/10.5153/sro.642 (acc. 15 January 2024).

Williams, R. 1980 *The Nuclear Power Decisions: British Policies, 1953–78* (London: Croom Helm).

Woodward, C. 1990 'The buildings for the Joint European Torus', *Proceedings of the Institution of Civil Engineers*, 88, issue 5, 855–73.

Wooldridge, T., and Druce, S. 2019 *Golden Egg or Poisoned Chalice: The Story of Nuclear Power in the UK* (Northampton: British Institute of Non-Destructive Testing).

World Nuclear News 2022 'Sellafield ends nuclear fuel reprocessing after 58 years', https://world-nuclear-news.org/Articles/Sellafield-ends-nuclear-fuel-reprocessing-after-58 (acc. 14 September 2023).

Wynn, H. 1994 *The RAF Strategic Nuclear Deterrent Forces: Their Origins, Roles and Deployment, 1946–1969. A Documentary History* (London: HMSO).

Young, S. A. 1967 'Research and development', *Nuclear Engineering* (July), 536–8.

Youngman, P. 1985 'Sizewell nuclear power stations', *Landscape Design*, 154 (April), 28–35.

Index

Page numbers in **bold** refer to figures; page numbers in *italics* refer to tables.